Axiomatization of the Theory of Relativity

Axiomatization of the Theory of Relativity

by HANS REICHENBACH

Translated and edited by Maria Reichenbach
Foreword by Wesley C. Salmon

University of California Press 1969
Berkeley and Los Angeles

University of California Press
Berkeley and Los Angeles, California
University of California Press, Ltd., London, England

First published in German under the title
Axiomatik der relativistischen Raum-Zeit-Lehre

Library of Congress Catalog Card Number: 68-21540
Designed by Jorn B. Jorgensen
Printed in the United States of America

Foreword to the English Edition

WESLEY C. SALMON

Any profound scientific revolution provides an irresistible philosophical challenge, and often the stimulus for fundamental philosophical advances. The history of thought gives ample evidence of this fact, and the twentieth century is no exception; indeed, it exhibits the leading case in point. This century has been the scene of unprecedented scientific innovation, which is no doubt responsible for the remarkable concomitant growth in philosophy of science. Contemporary philosophy of science is a highly developed technical discipline whose ramifications extend deep into our understanding of science and into philosophy in its broader aspects. Einstein's special and general theories of relativity constitute *the* major impetus for this development. That these theories have enormous philosophical import is a commonplace; it is a more difficult matter to specify the exact nature of this significance.

For two and a half millennia, the concepts of space and time have played a central role in the attempt to establish scientific knowledge of the physical world. Throughout this long history, these concepts have given rise to extremely perplexing *logical* and *epistemological* problems which do not automatically disappear with the advent of improved empirical or mathematical knowledge; indeed, they frequently *result from* the strictly scientific advances. In the beginning, for instance, when systematic geometrical knowledge was starting to emerge, Pythagoras, Zeno, Plato, and Aristotle grappled with just such conceptual problems. Centuries later, when Galileo and Newton had ushered in modern science, philosophers such as Descartes, Leibniz, Locke, and Kant (some of whom had themselves made scientific contributions of great importance) struggled with the epistemological problems of space and time issuing from that scientific revolution. And when the nineteenth century produced the non-Euclidean geometries, philosophers like Poincaré and Russell attempted to cope with the attendant conceptual difficulties.

The appearance in 1905 of Einstein's special theory of relativity made it evident that our spatial and temporal concepts would again have to undergo fundamental revision. In some cases, relative concepts (e.g., simultaneity) would have to replace absolute ones, while in others, absolute concepts (e.g., space-time interval) would replace relative ones. Space and time would be more intimately related than had previously been realized. But *much more basically*, the very meanings of such concepts as length, temporal interval, and simultaneity would have to be reexamined and revised. It turned out that conventions or definitions were required whose need had not previously been anticipated; moreover, the need for these conventions depended not only upon logical and semantical considerations, but also upon the physical facts of nature. As a result, a theretofore largely unsuspected degree of conventionality emerged in physics, the paradigm of *objective* science, and the conventionality was itself a consequence of objective facts. This conventionality had been detected to some extent by Poincaré prior to special relativity, but its full measure could only be appreciated as a result of Einstein's work. The primary philosophical demand of the theory of relativity was to understand, thoroughly and in technical detail, the role of convention, the role of empirical fact, and the intricate relations between them in a unified body of physical knowledge.

These problems pose a severe epistemological challenge to our understanding of the nature and foundations of the knowledge claims embodied in such physical theories as special or general relativity. It was to these philosophical questions that Hans Reichenbach directed his work on space and time. Having been trained in the Kantian tradition, he soon discovered what seemed to him an outright incompatibility between Einstein's physics and Kant's synthetic a priori conception of space, time, and causality. In his first book, *Relativitätstheorie und Erkenntnis Apriori* (1920),[1] he argued the case in detail. Eight years later, his *Philosophie der Raum-Zeit-Lehre*[2] set forth a comprehensive philosophical treatment of the problems of space and time that arise out of the discovery of non-Euclidean geometries and Einsteinian relativity.

The importance of this latter work can hardly be overrated. In it, Reichenbach offers an extremely sophisticated analysis of the status of geometry, taking into account not only the *logical* fact that consistent alternative formal systems of geometry exist, but also the *epistemological* and *experiential* considerations that are involved in the use of such systems for formulating spatial knowledge of the physical world. His analysis thus constitutes a refutation of Kant's views on the nature of geometry, and those of neo-Kantians who tried to reconcile Kant's basic philosophical outlook with the existence of the non-Euclidean geometries. In connection with geometry, he discusses the conventional status of spatial congruence; going on to temporal concepts, he deals with the conventional character of temporal uniformity and distant simultaneity. He reconciles the conventionality of all these concepts with the objective status of physical knowledge through the careful elaboration of his notions of coordinative definitions and equivalent descriptions. This work constitutes an epoch-making event in the philosophy of space and time, and no subsequent serious work in this area of philosophy of science can afford to ignore it. Moreover, the philosophical methods

[1] English translation: *The Theory of Relativity and A Priori Knowledge* (Berkeley and Los Angeles: University of California Press, 1965).

[2] English translation: *The Philosophy of Space and Time* (New York: Dover Publications, Inc., 1957).

and concepts that emerge have broad applications in other areas of philosophy of science, whether or not they directly involve space and time. This remarkable book may well be the finest single work in philosophy of science in the first half of the twentieth century—a model of philosophy of science at its best.[3]

The present book is a translation of *Axiomatik der relativistischen Raum-Zeit-Lehre*, which first appeared in 1924, midway between the German originals of *The Theory of Relativity and A Priori Knowledge* and *The Philosophy of Space and Time*; it constitutes a necessary prerequisite to a full understanding of the latter work. Reichenbach's approach to philosophy of science rejects the notion that the proper business of philosophy is to use the results of a scientific theory "as a whole" to construct a metaphysical system, a synoptic world view, or pious generalities about "the nature of things." Instead, he insists that the central epistemological questions can be fruitfully investigated only by careful and precise logical dissection of the theory. This is the reason that an axiomatization of it is crucial. In order to resolve such basic epistemological issues as the relations between fact and convention in physical theory, it is necessary to set out, one by one and in detail, the separate factual claims upon which the theory rests, and the conventions or definitions that are required to complete the statement of the theory. By investigating the logical consequences of these axioms and definitions, one sees how the various assertions of the theory arise from the particular axioms and definitions employed. Readers of *The Philosophy of Space and Time* will recall that when Reichenbach gets down to the absolutely fundamental task of construction of the space-time metric (§ 27), he relies entirely

[3] Although Reichenbach made contributions of greatest significance to other areas of philosophy of science—e.g., probability and induction, the philosophy of quantum mechanics—his interest in space and time never flagged. His last work, published posthumously, is *The Direction of Time* (Berkeley and Los Angeles: University of California Press, 1956).

upon the results of his axiomatization (while also correcting a few errors in it).

As Reichenbach clearly states at the outset of his work, axiomatization can be undertaken for a variety of reasons, and executed in different ways. If one is chiefly interested in mathematical elegance, the aim is to find a minimum set of maximally concise axioms from which rich consequences follow by ingenious derivations. Reichenbach's purpose in axiomatizing relativity theory is entirely different. Instead of striving for mathematical elegance, he is searching for *physical* clarity and insight, with the result that he takes physically elementary assumptions (in a sense he explains) rather than highly abstract or deep ones as his point of departure. He provides a relatively large number of empirically separable assumptions, so that the various physical aspects of the theory can be seen in relation to one another, *and* in relation to those of the classical theory which the newer theory supplants. He is *not* trying to present a minimal set of comprehensive statements as an exercise in mathematical frugality. Critics of Reichenbach's axiomatization, from Hermann Weyl[4] to Patrick Suppes,[5] have, I think, mistaken Reichenbach's fundamental purpose, and the value of his approach.

The import of Reichenbach's kind of philosophical analysis cuts two ways. For the individual who wishes to understand relativity as a physical theory, it offers a far more detailed analysis of the logical and conceptual foundations than is to be

---

[4] Hermann Weyl reviewed Reichenbach's *Axiomatik der relativistischen Raum-Zeit-Lehre* in *Deutsche Literaturzeitung*, Nov. 1924, col. 2122. Reichenbach replied to this and other criticisms in "Über die physikalischen Konsequenzen der relativistischen Axiomatik," *Zeitschrift für Physik*, XXXIV (1925), 32–48; see especially § II, 37–38. Weyl's review does not contain detailed reasons for his unfavorable opinion, but Weyl has reported that these reasons are given in detail in his lectures delivered in the Winter Term of 1930–31 at the University of Göttingen; a bound typescript of these lectures, under the title "Axiomatik," is in the library of the Institute for Advanced Study at Princeton.

[5] "The Desirability of Formalization in Science," *Journal of Philosophy*, LXV (1968), 651–664.

found in most physical treatises. In this connection, we should recall, Einstein himself was most emphatic about the absolute indispensability of logical and conceptual clarity in the enunciation of his theories. For the philosopher, with general epistemological concerns, his analysis provides a detailed examination of scientific knowledge in one of the most perfect forms yet to be created. Lessons are to be learned from this example which have profound implications for all aspects of human knowledge.

★ ★ ★

Bibliographical note. The reader of Reichenbach's *Axiomatization of the Theory of Relativity* may be interested in comparing other axiomatic approaches. The following list of references, though far from exhaustive, will hopefully prove to be suggestive.

Basri, Saul A. *A Deductive Theory of Space and Time*. Amsterdam: North-Holland Publishing Co., 1966.

Carathéodory, C. "Zur Axiomatik der Speziellen Relativitätstheorie," *Sitzungsberichte der Preussischen Akademie der Wissenschaften, Physikalische und Mathematische Klasse*, 1924. Pp. 12–27.

Carnap, Rudolf. *Introduction to Symbolic Logic and its Applications*. New York: Dover Publications, Inc., 1958. Chap. G.

Henkin, Leon, Patrick Suppes, and Alfred Tarski, eds. *The Axiomatic Method*. Amsterdam: North-Holland Publishing Co., 1959. Articles by Patrick Suppes, A. G. Walker, and Yoshio Ueno.

Robb, A. A. *A Theory of Time and Space*. London: Cambridge University Press, 1914. Also, *Geometry of Time and Space*. London: Cambridge University Press, 1936.

Schnell, K. *Eine Topologie der Zeit in logistischer Darstellung*. Dissertation, Münster in Westfalen, 1938.

Whyte, L. L. "Light Signal Kinematics," *British Journal for the Philosophy of Science*, IV (1953–54), 160–161.

Pittsburgh
January 1969

Preface

Seldom have philosophical tools been used in the discussion of a physical theory to the extent they have been used in discussing Einstein's theory of relativity. Not only did philosophers study this theory; physicists, too, felt compelled to base their presentations and analyses of the theory upon philosophical concepts. This fact is neither an accident nor the mere effect of a widely observable wave of philosophical interest in the foundations of the exact sciences. The theory of relativity simply invites epistemological foundations; it is not even intelligible without a clear understanding of the basic methods of physical science. A philosophical critique of these methods enabled Einstein to resolve long-standing difficulties in the explanation of physical facts and to introduce a new and fruitful conceptual framework.

Yet a survey of the enormous literature on the theory of relativity shows amazing differences of opinion in the evaluation of these epistemological foundations. I am not concerned with the numerous naïve contributions whose authors were not even aware of the philosophical problems and, therefore, became primitive adversaries of the theory—attacking it for its unintelligibility, counterintuitiveness, and internal contradictions. This sort of criticism did not help the theory: the continual repetition of banal objections whose futility was clearly recognized by the initiated diverted attention from the genuine epistemological problems of the theory and caused some scientists familiar with the theory to tire of and turn away from the epistemological discussion. Instead, I have in mind the—equally numerous—informed publications presenting accurately the physical and mathematical construction of the theory, and also the few purely philosophical studies concerned with a logical analysis of the content of the theory. However valuable the results of some of these works are, even in their contributions to the epistemological evaluation of the theory of relativity, primarily they are psychologically attractive presentations of the theory and philosophical evaluations of its position within the total framework of empirical knowledge rather than detailed epistemological analyses of the theory concerned with its internal conceptual structure. This is the reason the fundamental concepts of the theory are still somewhat vague. Concepts such as absolute time, time order, clock, rigid body, coincidence, natural metric, and constancy of the velocity of light are used without further clarification; "empirical result" and "arbitrary definition" are mentioned, yet it is not possible to say in specific instances what is the result of experience and what are conceptual additions. This imprecision appears negligible to the mathematical physicist, because his major aim is to escape as rapidly as possible from the philosophical realm into the domain of positive results, and the philosopher often does not pay attention to such imprecision, because he is more interested in incorporating the theory into a philosophical system than in an epistemological analysis of the theory itself. But the effect of this imprecision is the wide variety of opinions with which the consequences of the theory are judged.

The present study has been undertaken with these ideas in mind. It is not concerned with the position of the theory of relativity relative to philosophical systems. Neither does it aim at positive results that would enrich the domain of physics. Its goal is the construction of the relativistic theory of space and time with epistemological rigor. The *axiomatic method* is the only method that will reveal the logical structure of the theory with perfect clarity. The distinction between axiom and definition leads to the separation of empirical content from arbitrary concept formation, and the derivations of particular statements clearly reveal the empirical and the logical components of every assertion. The rigor of demonstration leads at the same time to exhaustive formulations of the assumptions and automatically excludes the formation of vague concepts.†

The present investigation was begun in the fall of 1920 and essentially concluded in March 1923. The author gave a brief summary of the results in two parts at the Physics Congresses in Jena in 1921 and in Leipzig in 1922; an abstract of the first paper was published in *Phys. Zeitschrift*, vol. 22 (1921), p. 683. This work, however, does not yet contain the axioms in their final form and numbering. Publication was postponed because of the difficult economic situation, and I am indebted to the *Württembergischen Gesellschaft zur Förderung der Wissenschaften* for their help in finally publishing the book. I am grateful to the publisher, Friedr. Vieweg & Sohn Akt.-Ges., for publishing this book at a time so unfavorable to the publication of scientific works. I should like to thank Dr. K. Fladt, Vaihingen, for his friendly assistance with the editing of the text and the reading of some of the proofs.

I hope that this investigation which is restricted to a special problem will at the same time stimulate further epistemological research in physics. It is realized more and more in philosophical circles that an exhaustive theory of empirical knowledge can be formulated only in close connection with actual scientific theories. It requires mathematical analysis rather than cursory inspection to discern the logical structure of empirical knowledge. There are already many physicists and mathematicians who expect the construction of an exact theory of knowledge to result from the logical analysis of science, a construction that cannot be provided either by the a priori–deductive or by the historically oriented descriptive

† See editor's reference notes, p. 202.

method of philosophy. The ultimate goal is still too far away for a general survey of results, and therefore it appears more urgent to seek solutions to individual problems rather than to attempt to formulate a comprehensive theory of knowledge.†

Hans Reichenbach

Stuttgart
March 1924

† See editor's reference notes, p. 202.

Acknowledgments

I am most grateful for the generous support I have received from the National Science Foundation, and I am also greatly indebted to the San Fernando Valley State College Foundation for administering the grant and for providing the opportunity to carry out this project. I further wish to thank Professor Wesley C. Salmon for writing a foreword to the English edition. Special thanks go to my friends, Hilary and Ruth Anna Putnam, Wesley C. Salmon, Adolf Grünbaum, and Sidney Trivus, who patiently read the manuscript and made valuable corrections. For their assistance in preparing the manuscript for print, I want to thank the editors of the University of California Press, Geoffrey Ashton and Vincent Ryan, and my students, Carol Graham and Jerry Wood.

M. R.

Los Angeles
April 1969

Contents

INTRODUCTION

FIRST PART: SPECIAL THEORY OF RELATIVITY

Chapter II. *Critical Considerations*

SECOND PART:
GENERAL THEORY OF RELATIVITY

Chapter III. *The Axioms and the Construction of the Metric*

Chapter IV. *Integral Properties*

Abbreviations
of the titles of works by
Hans Reichenbach referred to in editor's notes

AR — *Axiomatization of the Theory of Relativity*. Berkeley and Los Angeles: University of California Press, 1969.

DT — *The Direction of Time*. Berkeley and Los Angeles: University of California Press, 1956.

EP — *Experience and Prediction*. Chicago: University of Chicago Press, 1938.

MPS — *Modern Philosophy of Science*. London: Routledge & Kegan Paul, 1959. (Contains translations of articles originally published earlier than or simultaneously with *AR*.)

PST — *The Philosophy of Space and Time*. New York: Dover Publications, Inc., 1958. (Translation of German *Philosophie der Raum-Zeit-Lehre*, Berlin: Walter de Gruyter & Co., 1928.)

Introduction

Introduction

§ 1. *The Logical Status of the Axioms*

The value of an axiomatic exposition consists in summarizing the content of a scientific theory in a small number of statements. Any evaluation of the theory may then be limited to an evaluation of the axioms, because every statement of the theory is implicitly contained in the axioms. The *axiomatic method* has most frequently been applied in mathematics. It has become customary to reduce a controversy about the logical status of mathematics to a controversy about the logical status of the axioms. Nowadays one can hardly speak of a controversy any longer. The problem of the axioms of mathematics was solved by the discovery that they are definitions, that is, arbitrary stipulations which are neither true nor false, and that only the logical properties of a system—its consistency, independence, uniqueness, and completeness—can be subjects of critical investigation. The distinction between implicit and explicit definitions had to be introduced; the latter define a concept by means of previously defined concepts, whereas the former explain a number of concepts by giving a series of interrelations without showing a certain concept to be a direct (i.e., an explicit) function of other concepts.[1]

[1] D. Hilbert, in his *Foundations of Geometry* (The Open Court Publishing Co., 1921), calls implicit definitions axioms and explicit definitions explanations.

There is, however, a fundamental difference between physics and mathematics. Physical statements are more than mere consequences of arbitrary definitions; they are supposed to describe the real world. This means in the final analysis that phenomena that are directly observable can be derived from them. The ultimate criterion for the correctness of a physical theory is therefore always the question of whether such a phenomenon does or does not occur. "Truth" and "falsehood" have entirely different meanings in physics and in mathematics; to judge that a statement in physics is true is not a logical judgment but a judgment concerning the occurrence or nonoccurrence of sense perceptions. To the physicist the question of truth is the most interesting one, for if his theory is true, he can call it in a certain sense a description of reality.

The axiomatic exposition of a physical theory is at the outset subject to the same laws as that of a mathematical theory: it must satisfy the logical requirements of consistency, independence, uniqueness, and completeness. Yet since the physical axioms also contain the whole theory implicitly, they must themselves be justified: they must not be arbitrary but true. "True" refers again to a factual judgment ultimately tested by perception.

It is not easy to arrive at such a judgment with respect to the *axioms* of a theory. Usually the axioms, representing higher levels of abstraction, are quite remote from direct sense perception. It is frequently much easier to confirm a particular statement of a theory than its axioms. Therefore, only an indirect confirmation of the axioms will be possible. The confirmation of derived statements by means of experiments makes it possible to infer the truth of their presuppositions with a certain probability. We might think of the content of a theory as summarized in a variational principle; this principle can never be the direct object of experimentation, and yet, depending on the confirmation of its consequences, it may be called true or false with a certain degree of probability.

In order to avoid this difficulty, it is advantageous to approach the axiomatization in a different fashion. It is possible to start with the observable facts and to end with the abstract conceptualization. A certain loss in formal elegance will be balanced by logical clarity. The empirical character of the axioms is immediately evident, and it is easy to see what consequences follow from their respective confirmations or disconfirmations. Such a *constructive* axiomatization is more in line with physics than is a *deductive* one, because it serves to carry out the primary aim of physics, the description of the physical world.

Unfortunately, a new difficulty arises. Every factual statement, even the simplest one, contains more than an immediate perceptual experience; it is already an interpretation and therefore itself a theory. To infer the existence of a light source from the perception of light, or the existence of an external mass from the perception of pressure, is an extrapolation which only seems to be a matter of course, because its consistency has been tested innumerable times. It will be impossible, however, to make the content of a sense perception an axiom of a theory; it will always be necessary to extrapolate to a certain extent in order to obtain a factual statement from which consequences will be derivable. At least, extrapolations of the kind used in everyday life will be necessary. We shall not say, "under these circumstances I have the sensation *bright*," but "under these circumstances a lamp is burning." Even that is not sufficient. We shall have to make use of the scientific theory itself in order to interpret the indications of our measuring instruments. Thus we shall not say, "a pointer is moving," but "the electric current is increasing." The most elementary factual statements, therefore, contain some measure of theory.

In addition to theory, the factual statements presuppose epistemological principles. One of them is the assumption that the experiment, if subsequently repeated, will always yield the same result, the assumption of causality. Furthermore, the principle of induction is presupposed; for instance, certain values of measurement will be connected by a simple curve and this curve will be called an empirical law. An elementary statement therefore contains a certain amount of epistemology.

Will it still be advantageous, under such circumstances, to start an axiomatization with so-called empirical facts? Are we permitted to consider such particular facts to be more certain than their confirming theory? Does there exist any confirmation other than that of the theory as a whole? Many answer this question negatively, but they are mistaken. There is a way out which is peculiar to all factual knowledge and which rests on the possibility of approximation. Let us assume that the theory by means of which we explain a certain fact is false and is to be replaced by a different one. It is nevertheless possible that the new theory, when used for the interpretation of this one fact, makes hardly any difference quantitatively, whereas it leads to considerable changes with respect to other assertions. For example, whatever assumptions may be made about the motion of light in strong gravitational fields, its motion within the telescope will in practice be assumed to be always the same. In this sense, some facts are more elementary than others, namely, those whose interpretation within certain limits does not depend upon theoretical conceptions. This is why the facts of daily life appear so certain to us. Because of their imprecision they are invariant relative to all scientific theories.

Accordingly, this investigation starts with *elementary facts* as *axioms*; all are facts whose interpretation can be derived from certain experiments by means of simple theoretical considerations. Such a treatment (or separation) presents itself all the more readily, since the theory whose foundation is the subject of this book—Einstein's theory of relativity—constitutes an innovation in physics, and the elementary facts can be chosen in such a way as not to presuppose the new theory for their interpretation in connection with the experiments on which these facts are based. The new theory has merely to satisfy the requirement that it will not result in a practically noticeable quantitative difference when applied to these elementary facts. This requirement is satisfied by the theory of relativity. For this reason all axioms of our presentation have been chosen in such a way that they can be derived from the experiments by means of *pre-relativistic* optics and mechanics. All are facts that can be tested without the use of the theory of relativity. In particular, they are all formulated without the use of the concept of simultaneity at distant points.[2] The possibility of such a separation of elementary facts concerning the theory of relativity seems to be an especially valuable result of the axiomatization. This separation refutes at once the view that the theory of relativity asserts unimaginable facts. *The particular factual statements of the theory of relativity can all be grasped by means of pre-relativistic conceptions; only their combination within the conceptual system of a theory is new.*

Some of these axioms—such as II,4; IV,1; VI; IX—have been formulated in a somewhat different manner; in order to make them more intuitive, they have been conceived as "thought experiments." In the given form they would not be directly experimentally tested. But experiments from which these axioms are derivable can easily be indicated, again by means of pre-relativistic physics alone. Therefore, they do not violate the principle of using elementary facts. Such a schematic treatment has the advantage of making the empirical character of the assumptions immediately clear.

Using the concept of elementary fact, we may therefore say that the axioms of our presentation are *empirical propositions.* We shall show how Einstein's theory of space and time can be derived from them. In this construction it will become obvious which parts are empirical principles and which parts are logical additions. We shall formulate the logical part in the definitions.

§ 2. *The Logical Status of the Definitions*

Definitions are arbitrary; they are neither true nor false. They are merely to be analyzed with respect to their logical properties,

[2] Axiom IX is an exception; it has been formulated differently in order to make it more intuitive. But this axiom can be replaced by axiom *D* (§ 24) which does *not* contain the concept of simultaneity.

their uniqueness, consistency, and, under certain conditions, their simplicity. It is characteristic of the axiomatization of physics compared to that of mathematics that there exists such a distinction between axioms and definitions; an essential task of the axiomatization consists in tracing this distinction within the theoretical system.

However, even definitions in physics are different from definitions in mathematics. The mathematical definition is a *conceptual definition*, that is, it clarifies the meaning of a concept by means of other concepts. The physical definition takes the meaning of the concept for granted and coordinates to it a physical thing; it is a *coordinative definition*. Physical definitions, therefore, consist in the coordination of a mathematical definition to a "piece of reality"; one might call them *real definitions*. Thus the concept of the unit of length is a mathematical one; it asserts that a certain particular interval is to serve as a [standard of] comparison for all other intervals. From this nothing can be inferred, however, as to which physical interval is to serve as the unit of length. The latter is first accomplished by the coordinative definition which designates the Paris standard meter as the unit of length. In this physical definition, the mathematical definition of the concept is presupposed.

The difficulty concerning coordinative definitions is similar to that concerning elementary facts: the physical thing that is coordinated is not an immediate perceptual experience but must be constructed from such an experience by means of an interpretation. If I establish the coordinative definition "a light ray is a straight line," then the coordinated physical thing, the light ray, is a construction going beyond perception. We eliminate this difficulty in the same way as we did above: for elementary interpretations we use coordinative definitions whose degree of precision is not important and which, in particular, do not make use of relativistic definitions. This device leads to a restriction in the arbitrariness of coordinative definitions: they must not contradict elementary definitions. For instance, a cube must not be defined as a straight line.

Such a definition would require a reorganization of the elementary theory, even the "theory of everyday life."

The *simultaneity of spatially separated events* requires a coordinative definition. This example serves well to point out the arbitrariness of the coordinative definition and its connection with the axioms. The concept of simultaneity is defined mathematically as follows. *Two events at distant places are simultaneous if the time scales at the respective places indicate the same time values for these events.* In order to produce this simultaneity physically, a rule is needed to determine the numerical value of a time interval and to designate, for example, light signals for the comparison of time intervals. Such a rule is a coordinative definition. Just like conceptual definitions, coordinative definitions are arbitrary, and "truth" and "falsehood" are not applicable to them. Certain requirements may be imposed on the definition of the concept of simultaneity, for instance, that the time values on the clocks should satisfy certain requirements of causality (cf. theorem 9); the selection of the coordinative definition will then be restricted. Which requirements are necessary in order to make the coordination *unique* cannot be discovered a priori; these requirements can only be derived from the facts laid down in the axioms. The requirement of causality is not sufficient for the uniqueness of simultaneity if axiom II, 5 is valid; it is sufficient, however, when axioms *F, 1* and *F, 2* (§ 25) are valid. Furthermore, it cannot be known a priori whether the definitions are consistent; this result, too, follows only from the axioms. A relevant example is the consistency of definitions 10 and 16 which presupposes axiom V. It can be seen that the logical structure of a theory does not depend upon the definitions alone but also upon the axioms. It is the task of the theoretician to choose his definitions in such a way that, given the axioms, they lead to simple logical relations. This requirement is satisfied to a high degree in the theory of relativity, and in this property lies the attractiveness of this theory. On the other hand, the theory of relativity, in opposition to the classical theory of space and time, asserts that in view of the elementary facts certain concepts are no longer applicable, that is, they are no longer unique or consistent (cf. § § 22 to 27). In this sense the facts necessitate new conceptual systems without singling out any specific one of them.

Since the necessity of a coordinative definition for the application of the concept of simultaneity has often been overlooked, I should like to add a few remarks. We are speaking of simultaneity at different space points, not at the same place. It is impossible to *know* whether two events at different space points are simultaneous. Let these space points be P_1 and P_2, and let the events be E_1 and E_2. In order to compare the times of these events, a causal chain, a signal, would have to travel from E_2 at P_2 to P_1 which would arrive at P_1 later than E_1 at the time t_1. By measuring the time difference $t_1 - E_1$ at P_1 and keeping the velocity of the signal in mind, one could infer at what time the signal had left P_2; only in this way is a time comparison of E_1 and E_2 possible. The measurement of simultaneity presupposes the knowledge of a velocity. On the other hand, the measurement of a velocity presupposes the knowledge of the simultaneity, because time measurements at two different places are required. At one space point only the sum of the time intervals for a round trip, that is, an average velocity, can be measured—from which nothing can be inferred concerning the velocity in *one* direction. The attempt to decide about simultaneity by *measurements* leads to a circular argument. Simultaneity of distant events therefore can only be *defined*, and all measurements will yield only that simultaneity which has previously been determined by definition. Even if one tries to circumvent the use of a signal for the determination of simultaneity and determines it by the transport of clocks (§ 23), one does not change the definitional character of such a procedure. All measurements that are made relative to such a synchronization reflect the synchronization in their numerical values and can never be a test of its truth. From these values one can infer only that those events are simultaneous which had previously been defined as simultaneous by the rule. A different rule would therefore not lead to contradictions. I shall call this arbitrariness the *epistemological relativity of simultaneity*. It says that any definition of simultaneity is suitable for a description of the physical world. Restrictions result only from additional requirements such as those mentioned above. Einstein's principle of relativity is even more comprehensive. It says in the special theory that the use of a certain definition of simultaneity (def. 8) for uniformly moving systems leads to a complete isomorphism of all metrical relations so that the laws of nature have the same form for all such systems (principle of special relativity). In the general theory the principle says that given any definition of simultaneity, the laws of nature retain the same form no matter what metric is used, provided it agrees with the normal metric of the special theory in infinitesimal domains (principle of general covariance). These two principles assert the *physical relativity of simultaneity*. They contain empirically testable statements about the physical world. These statements, insofar as they concern metrical relations of space and time, are contained in the subsequent axioms. The fruitfulness of Einstein's physical theory depends, of course, on this more comprehensive empirical relativity. Its logical possibility, however, rests upon Einstein's discovery of the epistemological relativity of simultaneity.

The preference for a specific definition of simultaneity in the special theory is due to considerations of *simplicity*. Since the general theory of relativity is based in infinitesimal domains on the metric of the special theory, simplicity enters also into the metrical relations of the general theory, which, in spite of the complete arbitrariness of the coordinate systems, possesses a specific metric of the space-time manifold due to the specification of the respective $g_{\mu\nu}$. This simplicity has nothing to do with the truth of the theory. The truth of the axioms decides the empirical truth, and every theory compatible with them which does not add new empirical assumptions is equally true. Einstein's simultaneity according to definition 8 is not "truer" than any other because it leads to simpler relations;[3] any other definition is equally justified though it may lead to more complicated relations such as describing causal processes in negative time. This equivalence of descriptions is the essence of general relativity. The principle of simplicity in physics has a twofold nature.† First, it requires that concepts should be constructed as simply as is compatible with the facts. In this respect, of which the previous example is an instance, simplicity does not assert anything about the relation of the theory to reality, but contains merely a practical rule; the only issue is economy. The simplest description can always be transformed into a more complicated one, and thus the whole range of descriptions is accessible. This *descriptive simplicity*, therefore, does not involve any epistemological problems. There exists another principle of simplicity in connection with inductive inferences; for instance, the simplest curve is fitted to a set of points that represent measurements. This *inductive simplicity* is a principle of probability; it contains an important assertion about the physical world, that is, the statistical assertion that in most instances points representing future measurements will be located on the same simplest curve. This statement is either true or false. Whole theories can be compared with one another from the point of view of inductive simplicity, and (with a certain probability) one of them is to be regarded as true and the others as false. An analysis of inductive simplicity belongs to the theory of probability. A confusion of these two principles of simplicity is the reason for many pseudoproblems—for example, the questions why the simplest theory is the true one or whether Euclidean geometry should be called true because of its simplicity. In the realm of space and time all simplicity properties of the theory of relativity belong to descriptive simplicity and have no bearing upon the truth of the theory, which depends solely

[3] That this simultaneity leads to simpler relations is an assertion that may be true or false. But this simultaneity itself is not therefore true or false—only logically preferable because simpler.

† See editor's reference notes, p. 202.

on the axioms. Only in the relativistic theory of gravitation does inductive simplicity play a role.

§ 3. *Provisional Remarks Concerning the Results of the Axiomatization*

The view that every spatial and temporal metric presupposes coordinative definitions has been generally accepted and is known as *conventionalism*. For spatial measurements it originated with Helmholtz and Poincaré; Einstein extended it to temporal measurements (simultaneity and uniformity). It is customary to use rigid bodies for spatial measurements, natural clocks for temporal measurements, and light signals for simultaneity only. It should be noted that for the purpose of employing a minimum of axioms in our presentation, the space-time metric of the special theory of relativity is defined by light signals alone, and therefore light signals are the basis not only for simultaneity but also for uniformity and equality of spatial distances.[4] This is possible without the use of the frequency of light, that is, without the use of an atomic clock (as a measure of uniformity) or the amplitude derived from it (as a measure of distance). The definition is given by means of geometrical optics alone. The result is a complete and unique light geometry that uses only light for the physical component in the coordinative definition, at least in finite domains. The other physical things used in the definition must be conceived of as infinitesimal compared to the spaces to be measured; they have no direct metrical significance.

The definition of the uniform state of motion in space succeeds only if the following criterion is accepted as admissible. The resulting metric must be without singularity at any space point. If such a criterion is rejected, there remain two light-geometrically equivalent classes of spatial reference systems which are in accelerated motion relative to each other, one of

---

[4] A hint that this is possible may be found in the Appendix of the fourth edition of H. Weyl, *Space, Time, Matter* (London: Methuen, 1922).

which, measured relative to the other, always experiences an internal expansion (or contraction) and therefore is not rigid. The characterization of one of the two classes as rigid and moving uniformly is accomplished by means of physical things, either by a rigid rod or a natural clock. These physical things are defined in definitions 18 and 19; a thorough analysis of this question will be found in § 16. Nevertheless, the result remains that for each of the two classes of systems the space-time metric is defined by light alone. Light geometry thus leaves the class ambiguous, but defines all relations uniquely within each class.

We shall state those properties of light which are necessary for the construction of light geometry in a special class of axioms, the *light axioms* (I–V). They do not refer to any relations between light and material metrical bodies. It is interesting that there are such axioms, that the mere motion of light is connected with ascertainable facts that might conceivably be different.

After we have developed the *light geometry*, we shall develop the *matter geometry* and strictly define the physical things used for this purpose, namely, rigid rods and natural clocks. Einstein's idea can be formulated as meaning that light geometry and matter geometry are identical. We shall present this idea in a second class of axioms, the *matter axioms* (VI–X).

This axiomatization, however, is not restricted to an analysis of metrical relations. It is preceded by an investigation of topological relations.† In this way conventionalism is extended to another level because the topological order will also be defined by coordinative definitions. Of prime importance is the time order; what is to be called *earlier* and *later* is defined by reference to *signals* (*causal chains*). Subsequently, the *topological problem of simultaneity* will be analyzed, that is, the question of how far topological definitions restrict the arbitrariness of simultaneity will be considered. Later, *spatial neighborhood*

† See editor's reference notes, p. 202.

will be defined by means of signals. It is not quite so well known that these topological relations are also definitions, not empirical discoveries. That these relations cannot be known a priori is revealed by the fact that the definitions do not completely determine the topological structure, and that only additional axioms, or empirical facts, determine the ultimate structure (§ § 7 and 25). The following section, however, will show that sense perception alone does not decide this question either.

The most important result that could be derived from this construction was an explication of the concept of time. Modern epistemology has shown clearly that time is not a form of pure intuition, as Kant believed; the intuitive experience of time is merely the psychological source from which stems a construction of the conceptual scheme of time holding for the physical world. Only then will the nature of this time order be clearly understood. The physical world consists of causal chains,† which show certain structural relations that can be formulated as axioms. *Topological* and *metrical* axioms can be distinguished. Topological axioms are statements of the kind that all space points can be connected by causal chains, or that no causal chain is closed, etc. Metrical axioms are statements about certain topologically specified causal chains (first signals) which characterize these chains in terms of a certain simplicity. The structure of the causal chains laid down in these axioms makes it possible to give a simple one-dimensional order to all events. The topological axioms make it possible to define a unique, directional relation *later than* between any two events; within this order the metrical axioms permit a choice of metric distinguished by its simplicity. We possess a time order only because the structure of the causal chains admits such an order. Therefore, time describes an objective state of the physical world just like any other scheme of order—for instance, genealogical order. The controversy concerning the ideality or reality of time—the same pertains

† See editor's reference notes, p. 203.

to space—is due to the fact that the concept of time does not refer to a thing but to a state of things. Yet this state of things is just as real as any other state of things—for example, heat. On the other hand, the concept of time, just as any other concept, is something ideal. *Time is the order type of causal chains;* it has therefore its own physical laws whose axiomatic formulation describes its specific character.

§ 4. *Coincidence*

Coincidence is frequently described as an immediate experience of simultaneity. This assertion is not quite correct. Such an experience furnishes only subjective coincidence; the temporal order of events, however, is based on objective coincidence.†

Subjective coincidence occurs, for instance, when the astronomer observes the coincidence of the position of a star in the telescope with the ringing of a clock. This is a psychological phenomenon because the merging of the two events in a coincidence takes place in the mind of the observer. Strictly speaking, this subjective coincidence has no defined location. The analysis of this phenomenon belongs to the philosophical theory of perception. Physics takes the possibility of this psychological function for granted.

In addition, there are objective coincidences in which, independently of the observer, two events coincide. The mind merely records such a point event. Examples are furnished by the coincidences of a pointer with a line on a scale, and of a light ray with the tooth of a gear (Fizeau's measurement of the velocity of light) or with another light ray (interference). All of these are instances of space-time coincidences. It merely seems as if in the last two examples time were not involved; since these are stationary processes, what is perceived does not change, but time enters fundamentally in the phase. Even in borderline cases where two things coincide permanently, this permanent coincidence should be analyzed as a sequence of

---

† See editor's reference notes, p. 203.

successive momentary coincidences as soon as it is described with greater precision.

Both methods—the observation of subjective and of objective coincidences—are used in physics. But there is a great difference between them. With respect to the second kind, coincidence is inferred, not perceived. Perceptually speaking, a totality of qualities is given, such as dark and bright spots or sound impressions; from these qualities the objective coincidence of things is inferred. When two billiard balls collide with each other, we hear a characteristic noise; but that this experience *noise* constitutes a coincidence of balls is a logical construction. The expression "I hear the collision" is an abbreviated way of speaking about the complicated meshing of perception and conceptualization. Optical impressions are similar. We infer the objective coincidence of a pointer with a line on a scale from the structure of a certain perception; it is only an abbreviation to say, "I see the pointer above the line on the scale."

This analysis is important, because the theory of relativity uses coincidences as elementary facts, and this is usually justified by the claim that ultimately all observations are reducible to coincidences. But the coincidences that the relativistic theory of space and time employs are of the objective kind, not those immediate experiences of coincidence represented in subjective coincidence. It cannot be maintained, therefore, that the point events of the theory of relativity are ultimate facts. Only subjective coincidence has the character of the immediately given; subjective coincidence alone is independent of all interpretations and is a necessary condition for all physical observations. It occurs wherever an objective coincidence is ascertained. In order to interpret the perception of a certain structure in the sense of a certain objective coincidence, other subjective coincidences must be present. If we observe, for example, the objective coincidence of a pointer with a line on a scale, this interpretation of our perception is admissible only if other simultaneous perceptions guarantee that the pointer is a physical thing and not a mirror image projected upon the scale. This simultaneity is judged by means of subjective coincidence. It is true that even in the case of a mirror image one can still speak of a coincidence; it is the coincidence of a light ray with the scale. But what actually coincides here is always inferred. Such inferences are made by means of other coincidences which ultimately are subjective and contain a reference to time. The role of subjective coincidence is seen more clearly when the objective coincidence is not so simple. For instance, if interference lines are observed in the telescope, they are evidence for an objective coincidence only if a certain lamp is burning simultaneously that is regarded as the cause of the lines on the basis of theoretical considerations. Only the subjective coincidence of the perception of the lines with the perception of the burning lamp permits interpretation of the lines in the sense of an objective coincidence. Even if the lamp is not seen—only "known" to be burning—observation of subjective coincidence must occur; the perception of the lines must be supplemented by perception of the surrounding instruments and of the room, from which it can be inferred that no disturbance takes place and the lamp is burning. It is easily seen that the shifting of coincidences to objective levels presupposes other subjective coincidences. The advantage of such a shift consists in the result that the subjective coincidence might be quite imprecise, whereas the objective coincidence would be very accurately determined thereby.

The assertion that all physical observations are reducible to coincidences holds, therefore, only for subjective coincidence. But this epistemological fact is no justification for the reduction to coincidences which occurs in the theory of relativity, because this theory is based on objective coincidence. As in so many other cases, a subjective experience is only the psychological source for the construction of a concept by means of which we describe the physical world. One must be careful, therefore, not to transfer the evidential character of the subjective experience of coincidence to the use of the concept of coincidence for a description of objective physical events. This coordination of the concept of coincidence to the physical event has merely the certainty of all empirical knowledge and depends upon certain definitions. The decision to base the theory of space and time, nevertheless, upon objective coincidence as an elementary fact is justified by the consideration that objective coincidences are elementary facts, in the sense of § 1, which remain relatively invariant with respect to a great variety of interpretations. But that is all that can be advanced epistemologically for a preference of objective coincidences as elementary facts. There is no absolute necessity to reduce all order relations to objective coincidences, nor can it be asserted that the concept of coincidence will forever be an element of concept formation in physics.

The detachment of the concept of coincidence from perception, however, is inevitable for the construction of the external world. Only this method eliminates the observer and replaces subjective experiences by objective facts. Here begins at once the strictness of quantitative determination. Coincidence becomes quantitatively precise in the same way as the detachment of the concept of physical force from the experience of force by the human body substitutes a precise measurement of force for an intuitive estimate. Coincidences of physical events, however, are never points in the mathematical sense. But they are much more precisely determined than subjective coincidences. The "space point" of a subjective coincidence, if we may talk this way at all, is as large as the human body; the space point of an objective coincidence can be much smaller. Even here there is a hierarchy. There are objective coincidences that are "directly observable"; they are coincidences whose space points lie above the threshold of perception. We shall call them *objective coincidences of the first degree.* One usually thinks of them when speaking of coincidences as intuitively given elementary facts. Examples are furnished by the coincidence of a light ray with a cross wire, or of a pointer with a mark on a scale. Even they are merely inferred as objective coincidences, but for their interpretations concepts of daily life and simple theoretical principles suffice. *Objective coincidences of the second degree* exist in subatomic dimensions; they are no longer "directly observed." Their interpretations contain more complicated theories and describe what is perceived as a more remote effect of the occurring coincidence. Examples are furnished by the collision of two atoms, or the incidence of a light ray upon an electron. The inference often starts from facts which themselves are objective coincidences of the first degree—for instance, the observation of an ammeter pointer on a scale. Such conditions are not necessary, however; there are cases in which coincidences of the second degree are inferred from purely qualitative determinations. For example, the collision of ions is inferred from the occurrence of an incandescent layer in cathode-ray tubes. The space point of such an objective coincidence of the second degree is considerably smaller than the observed phenomenon.

It is noteworthy that the statement about the precision itself is made by means of an inference that is based on the discussed auxiliary concepts without leading to contradictions. We discover the imprecision of subjective coincidences compared to objective coincidences of the first degree by means of a theory founded upon objective coincidences of the first degree (say, a graphic measurement of time), and the imprecision of objective coincidences of the first degree compared to those of the second degree is revealed by theoretical considerations which presuppose the existence of objective coincidences of the second degree.

None of these kinds of coincidences must be regarded as infinitesimal structures. Even the objective coincidence of the second degree is generally conceived of as extending over a finite space-time region, and it is very questionable—and not decidable on theoretical considerations alone—whether or not we can extend the concept of coincidence arbitrarily to smaller and smaller regions—for instance, to those below the dimensions of an electron. The conception of coincidence as a mathematical point is always permissible for sufficiently large dimensions. We merely commit the mistake of all physical theories in working with mathematically strict concepts in instances where physical objects correspond to them only approximately—a mistake that vanishes below the limit of exactness of all empirical knowledge.

An order of coincidences is given in the same way as the coincidence itself. We observe the temporal order of two subjective coincidences; it is questionable whether in this instance there also exists a spatial order or something similar. We observe a spatial and temporal order of objective coincidences in the same way as we observe the coincidence itself. Logically speaking, this order is inferred, but is of the same degree of factualness as the coincidence itself. Thus "we see directly" whether a star is located between two wires. In a world in which we may regard objective coincidences of the first degree as mathematical points without committing major mistakes, the infinitesimal order of these points, and thereby their topology, is perceived in the same sense as is coincidence. This order is therefore also an elementary fact and "inferred" merely in the epistemological sense. However, the infinitesimal order of objective coincidences of the second degree is not perceivable; in this instance, any immediate relation to the observer is lacking.

If one bases the order of space and time on objective coincidences regarded as elementary facts, then the infinitesimal order may equally be assumed a given fact. Yet this order is logically much more complicated than the coincidence itself, and it will be better not to make use of this possibility, but to connect this order with the observed fact of coincidence by means of special rules. In this way, a space-time system is obtained which is constructed by means of a smaller amount of empirical knowledge. At the same time, the principles will be developed according to which the infinitesimal order of objective coincidences of the second degree is inferred, an order that is no longer "directly observable." Furthermore, the principles will be revealed according to which even the order of objective coincidences of the first degree is strictly determined, an order for whose determination "direct observation" constitutes an abbreviated procedure. This method explains at once why the defined order does not contradict the perceived order wherever a comparison is possible.

First Part

Special Theory of Relativity

Chapter One

The Axioms and the Construction of the Metric

§ 5. *The Concepts of Real Point and Signal*†

We shall begin with an intuitive example. Let us imagine an empty space in which material points are traveling at random, similar to molecules in a gas. At every point sits a small observer who is equipped with physical instruments. He can send forth light rays, reflect them, etc. The observers have the task of constructing a temporal and spatial order; they are told to determine temporal succession, to synchronize time for all material points, to determine when two points are at rest relative to each other, and to measure spatial distances. They are supposed to solve all these problems solely by means of measurements on the various material points without the use of rigid rods. Signals are the only connecting links at their disposal.

This fictional example provides an intuitive picture of our problem while indicating at the same time the limits of visualization. What cannot be presented in this manner is a space continuously filled with points traveling at various velocities. We are trying to construct the mathematical configuration called the space-time continuum on which all physical conceptions are based. The intuitive representation of the world has holes; it consists of individual material points and occasional signals. We know, however, that we can presuppose the

† See editor's reference notes, p. 203.

continuum as a limiting configuration and that this limiting configuration, as the epitome of the possible arrangements of all material points and events, describes the actual arrangements better and more accurately than our intuition. Compared to this limiting configuration, the intuitive world is only an approximation—but in this interpretation we may always use it as an auxiliary tool when we want to construct the limiting configuration.

We shall therefore introduce the auxiliary concept *real point* and understand by it a point at which material entities may be conceived to be at rest. The real point corresponds more or less to the mass point of classical mechanics, with the reservation that we shall not make use of mass. But the possibility of individual markings by means of a "piece of matter" remains. A *real point* is distinct from a *fictitious point*, because in the theory of relativity points can be imagined that travel faster than light but cannot be material. The real point need not be conceived of as material; it can also be defined by reference to other materially marked real points (e.g., "the midpoint between two materially marked real points"). Yet this distinction is not very important.

The theory does not lose in generality by the assumption of individually marked real points. All axioms and theorems can be translated into statements about world lines which are not empirically and individually given but are selected from the field according to certain principles. Only because we are trying to carry through an intuitive separation of space and time as best we can do we distinguish the real point of three-dimensional space from the point event of the four-dimensional space-time manifold.

If we say that signals are produced or reflected at a real point, we commit once more the mistake of extending possibilities existing for small regions to the infinitesimal. In spite of this inaccuracy, we may carry through such a construction as a fictitious basis. By characterizing mirrors, telescopes, etc., as infinitesimal, we mean mainly that these material things have no direct metrical function; they are not used as rigid bodies defining the relative rest of neighboring real points or even measuring distances. Only signals are to be used for all metrical and topological functions.

By "signal" we understand a physical process that travels from a real point P to another point P' and has the following property: if this event is marked at P, the mark can also be observed at P'.[5] The word "signal" pinpoints this very property because it means a transmission of signs. The expression "causal chain" is also frequently used in such instances. Of course, light is a signal, but the restriction to light signals will not be introduced before axiom III. Within the limits of axioms I and II any signal may be employed—for instance, sound signals or traveling projectiles.[6] In these axioms we shall not speak yet of the path of a signal—only of real points with which it coincides. For the time being we shall therefore not specify in any way the manner of propagation. We do not demand that a signal must travel directly from P to P'; should a projectile originating from P collide at P' with a mechanism that sends another projectile to P'', the result will be again a signal PP''. This means that signals can be combined to result again in a signal. We do not yet distinguish between *signal chains* and *signals*. Particularly with respect to the validity of axioms I and II, it is irrelevant whether a signal is produced at P or

---

[5] We shall not enter into a discussion of the epistemological problem concerning the concept of mark; for our purposes the concept of ordinary language suffices. This concept indicates already that the mark at P' is not exactly the same as the mark at P, but shares only certain fundamental features. The mark may be distorted (sounds over the telephone). We shall leave open the question of whether a reference to the concept of Gestalt is necessary.

[6] In earlier presentations I started with light signals. The validity of theorem 9 for any real events then follows from the fact that the light signal PP' arrives at P' earlier than any other signal departing simultaneously from P (limiting character of the velocity of light). K. Lewin drew my attention to the fact that all properties of signals can be formulated directly for any signal (*Zeitschr. f. Phys.*, vol. 13 [1923], p. 62). Starting from differently oriented investigations concerning genealogical sequences, Lewin (*Begriff der Genese* [Berlin: Springer, 1922]) arrived at a conception of time order similar to mine. See the footnote to p. 39 concerning the difference between my time order and his.

merely reflected at P. That signals exist, that we can produce them, send them to a given real point, combine them, and reflect them are elementary facts; axioms I and II contain everything concerning these facts that is necessary for the construction of the order of time.

§ 6. *Axioms of Time Order*

Let us first define the time order at a real point by means of signals. A signal S is produced at P and given a special mark (light rays may be marked by color). The coincidence of S and P is called (SP), and the noncoincidence is called $S(P)$. At P the following events are observed: $S(P)$, (SP), $S(P)$. Nothing is asserted yet about the temporal sequence of these three events, because we do not make use of the "direct perception" of time order. Let the signal be "recognized" at P', for instance, by means of the color attached to it at P. At P' the signal is provided with a second mark (for instance, by polarization or a weakening of the intensity of the light) and reflected back to P. Now the coincidence $(\bar{S}P)$ is observed. The bar in "$\bar{S}$" is to indicate that this event is qualitatively different from (SP). In (SP) the signal S has only one mark which originated at P; in $(\bar{S}P)$ the signal S has a second mark which originated at P'. Due to these two marks, it is possible to distinguish between (SP) and $(\bar{S}P)$. We then define $(\bar{S}P)$ as *temporally later* than (SP), symbolized by $(\bar{S}P) > (SP)$. *Temporally earlier* is defined as the converse of *temporally later*: if $(\bar{S}P) > (SP)$, then $(SP) < (\bar{S}P)$. The coincidence of two events E_1 and E_2 is symbolized by $E_1 = E_2$.

We shall call the event (SP) the *departure of the signal* from P, the event (SP') the *arrival*[7] at P', the event $(\bar{S}P)$ the *return* to P. We shall say, the signal travels from P to P' and back from P' to P. Correspondingly, we shall say, a signal travels from P via P' to P'', etc; in this case a qualitative mark on the

[7] This definition does not say anything concerning the time relation between (SP) and (SP') since it pertains merely to the order of time at one point.

signal must be observable at P'' which had been attached at P', etc.

This consideration shows that it is possible to define the order of time at one point by means of qualitative differences between two events without the use of a "direct perception" of the order of time. The definition is based upon the causal connection of the two events which characterizes one of them as the cause of the other. The resulting order of time is identical with the temporal succession experienced in human consciousness and, in addition, indicates that direction which is characterized by irreversible processes. Thus there are three definitions of the order of time; their compatibility must be considered an empirical fact. It may be questioned, however, whether the first definition is independent of the third; the use of a mark on the causal chain may involve an irreversible process.† We shall not discuss this question here.

Let us summarize our definitions and add those axioms which assert the possibility of a general order of time. We shall add the theorems in our presentation in such a way that a theorem never presupposes axioms that follow it in the exposition.

Definition 1. Of two events E_1 and E_2 happening at P, the event E_2 is called later than E_1 if a signal chain can be chosen in such a way that its departure coincides with E_1 and its return with E_2. In this case E_1 is called earlier than E_2.

Axiom I, 1. There is no signal chain such that its departure and its return coincide at P.

This axiom says that there are no temporally closed world lines. It follows that definition 1 cannot lead to a contradiction.

THEOREM 1. If $E_1 < E_2$, there exists no signal chain such that its departure coincides with E_2 and its return with E_1.

If there were such a signal chain, there would also exist a compound signal chain $E_1 E_2 E_1$ which would contradict axiom I,1. Theorem 1 and axiom I,1 are logically equivalent;

---

† See editor's reference notes, p. 203.

theorem 1 can also function as an axiom from which the content of axiom I, 1 can logically be derived.[8]

It is, of course, possible that a signal departing from A will return twice (in consequence of a reflection at two different points B, C). It is presupposed that these events will be recognized to be two different events; their order of time can be determined by the use of definition 1 together with a new signal.

Axiom I,2. For any two events E_1 and E_2 at P there exists always a signal whose departure coincides with E_1 (or E_2) and whose return coincides with E_2 (or E_1).

Axiom I,3. The events at P form a linear continuum.

If $E_1 < E_2$ and $E_2 < E_3$, there exists a signal chain $PP' \ldots PP'' \ldots P$ for E_1 and E_3, and therefore, according to definition 1, $E_1 < E_3$; because of axiom I, 1 (theorem 1) there does not exist a signal chain such that $E_1 > E_3$. The relation *later than* is therefore *transitive*. According to theorem 1, it is also *asymmetrical*, that is, *later* and *earlier* are mutually exclusive relations: if $E_1 < E_2$, then neither $E_1 > E_2$ nor $E_2 < E_1$.

Axiom I, 1 is called the axiom of order; axiom I, 2 is called the axiom of connection; and axiom I, 3 is called the axiom of power. From axioms I, 1–3 there follows theorem 2.

THEOREM 2. A one-to-one correspondence can be established between the real numbers and the temporally ordered events at one point.

In order to facilitate the conception of time order at one point, we shall assume the existence of a clock at P. By "clock" we shall understand any mechanism that coordinates the events at P in their time order to the real numbers; the metric remains arbitrary. By means of axioms I the time order at one point is determined only up to a monotone function. Since the clock is not infinitely small, it is more precise to speak of a

[8] I used this method in my earlier presentation. Cf. *Phys. Zeitschr.*, vol. 22 (1921), p. 683.

clock next to *P*. The comparison of time between *P* and the clock may be assumed to be made by means of direct observation. We are making use of the above-mentioned faculty of the observer to perceive directly the practically infinitesimal neighborhood relations. In § 8 we shall define the comparison of neighboring events and justify this concept.

The assumption of a neighboring clock is not necessary for the determination of time; its only function is to assist visualization.

§ 7. *Axioms of the Comparison of Time*

The axioms of the comparison of time pertain to the comparison of time at two real points. Nothing is assumed concerning the relative state of motion of these real points, since we have not yet defined the concept of relative rest. Just as in axiom I, the real points are arbitrary. A special problem of the comparison of time is the definition of "simultaneity," that is, the determination of those events at two real points which receive the same time number.

Axiom II,1. Given an arbitrary time t at a real point P, a signal can be sent forth at t so that it will arrive at a given real point P'.[9]

Axiom II,2. Given an arbitrary time t' at a real point P', a signal can be sent from a given real point P so that it will arrive at P' at the time t'.

These two axioms assert that all real points can always be connected. Furthermore, we can derive from them in combination with axioms I,1 to I,3 the following theorem.

THEOREM 3. There exists at *P'* a lower time limit *t'* for the arrival of the signals *PP'* which can be sent from *P* at a given

[9] There are singular cases in which this axiom does not hold; cf. the example on p. 80. Axiom II,2, however, holds without exception. This fact indicates a fundamental difference between the two axioms.

time t. t' is called the time of arrival of the *first signal* PP' which departs from P at the time t.[10]†

Proof. At P' a time t' is selected so that the signal $P'P$ will arrive at P at the time t. If it were possible to send a signal at the time t from P which would arrive at P' at a time $T' < t'$, then $tT't'$ would constitute a temporally closed signal chain. The same result would be obtained if a signal departing from P at a time later than t were to arrive at P' before t'. Since such an event is excluded by axiom I, 1, t' is a lower bound for the arrival of the signals departing from P at the time t or later than t. On the other hand, if $t_1' > t'$ is the time of arrival of a signal departing at t, then according to axioms I, 2 and I, 3 there exists a signal $t_1't_2'$ so that $t_2' > t_1'$ and t_2' appproximates t_1' to an arbitrary degree. Therefore, there exist signals $tt_1't_2'$ which lie arbitrarily close to tt_1'. The signals PP' are dense above t_1', and therefore either t' or t_1' must be a lower limit (in contrast to a bound) for the signals departing at t, or there must exist such a limit between t' and t_1'. Even if this limit itself is never reached by any signal (and therefore is a lower limit of arrival times rather than a minimum), we shall call this limit the arrival time of the first signal.

It must be left open whether the first signal is a signal at all; strictly speaking, only the times of departure and arrival of the first signal are defined. Nevertheless, we may speak of "the first signal" because we shall always use only the times of departure and arrival (so long as we do not use axiom III). It should be noted, however, that as a consequence of this indeterminateness, axiom I, 1 does not necessarily hold for first signals. Furthermore, first signals and first signal chains must be distinguished; a first signal chain $PP'P''$ is not necessarily identical with the first signal PP'' but will usually arrive later. This is also the case when the first signal itself is a signal.

Axioms II, 1 and II, 2 still permit discontinuities in the time transfer from P to P', as is shown in figures 1 and 2. These

---

[10] Note that the definition of "first signal" is given without the help of the concept of velocity.

† See editor's reference notes, p. 203.

figures display the ascending world lines of P and P'; the cross lines are first signals PP'. In figure 1, tt' is a first signal; tT' is not a first signal, because it arrives at P' later than t'. Yet none of the signals departing from P later than t arrives within the interval $t'T'$, and tT' is the lower limit defined by

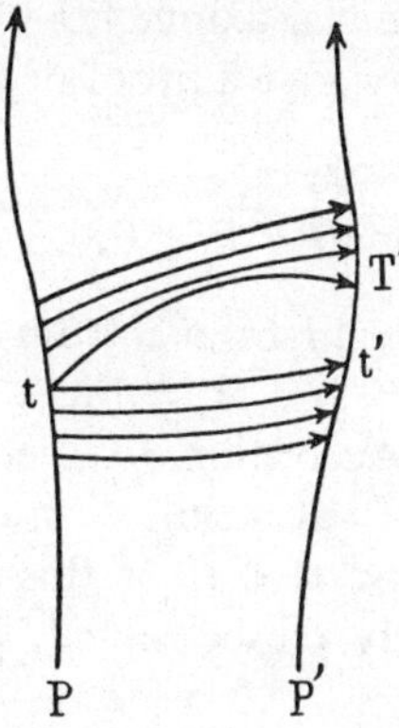

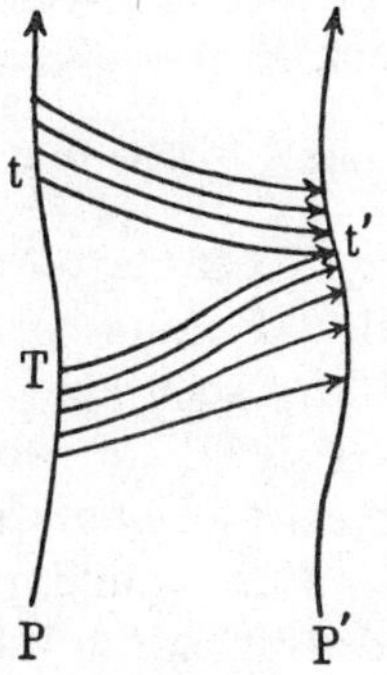

Fig. 1. Pertaining to axiom II, 3. Fig. 2. Pertaining to axiom II, 4.

the first signals departing later than t. The time interval $t'T'$ does not receive any first signals from P; the time point t is mapped upon the time interval $t'T'$. In figure 2, tt' is a first signal, but so is Tt' as well as all intervening signals to t'. Here the time interval tT is mapped upon the time point t'. Both cases are compatible with axioms I to II, 2; a crossing of the first signals is already excluded. We shall now exclude the cases of figures 1 and 2 (in this order) by the following two axioms.

Axiom II, 3. *If there is a signal tt' between the points P and P', then for each $t_1' > t'$ there exists a $t_1 > t$ such that there is a signal t_1t_1'.*

Axiom II, 4. *If there is a signal from P to P' at the time t, then for each time point $t_0 < t$ there is a signal leaving P at t_0 that will reach P' earlier than any signal departing at t.*

From this follows the next theorem.

THEOREM 4. If two first signals depart from P one after the other, they will arrive at P' one after the other, and a time interval at P will be mapped in a continuous fashion upon a time interval at P'.

Since first signals need not be signals, axiom I, 1 need not hold for them. However, we shall now give a proof of the next theorem.

THEOREM 5. For first signals the time $\overline{PP'P} \geqq 0$.[11]

Proof. Let t be the departure of the first signal from P, t' the arrival at P', and t_1 the return to P. If $t_1 < t$, there exists a signal $t't_2$ such that $t_1 < t_2 < t$. According to theorem 4, the first signal t_2P' would arrive at P' at a time $t_2' < t'$; there would exist a signal t_2t_3' such that $t_2' < t_3' < t'$. In this case the chain $t't_2t_3't'$ would be a temporally closed signal, a result excluded by axiom I, 1.

We shall now prove the most important theorem about time comparison.

THEOREM 6. It is possible to set all clocks in such a way that, for an arbitrary signal PP', its arrival at P' is later, or at least not earlier, than its departure from P.

Proof. We synchronize all clocks with a central clock O whose metric is arbitrary up to an arbitrary monotonic function. No selection takes place from the real points; rather for each arbitrarily moving real point P, the time is determined from O. This determination must be imagined to occur at every instant, because no mechanism can be established at P which would preserve a single synchronization forever. Every time point at P receives a signal from O, a process possible according to axiom II, 2; according to axiom II, 1 the signal can always be reflected back to O. We define the synchronism as follows.

[11] The bar in "$\overline{PP'P}$" means the time of the signal $PP'P$ measured at P.

Definition 2. Let a first signal, sent from O at time t_1 and reflected at P, return to O at time t_3. Then the instant of its arrival at P is to receive the time value

$$t_2 = t_1 + \varepsilon(t_3 - t_1) \qquad 0 < \varepsilon < 1 ,$$

where ε is an arbitrary factor which must have the same value for all points P.[12]

According to theorem 5, $t_3 - t_1 \geqq 0$; therefore, $t_2 \geqq t_1$, and theorem 6 is valid for the signal OP. Since $\varepsilon < 1$, $t_3 \geqq t_2$, and therefore theorem 6 is also valid for the signal PO. We must now show that this theorem holds for any arbitrary signal $P'P''$.

For this purpose we shall use figure 3. Let t_2' be the time of departure from P', and t_5'' the time of arrival at P''. According to definition 2, we obtain

$$t_2' = t_1 + \varepsilon(t_3 - t_1), \qquad t_5'' = t_4 + \varepsilon(t_6 - t_4) .$$

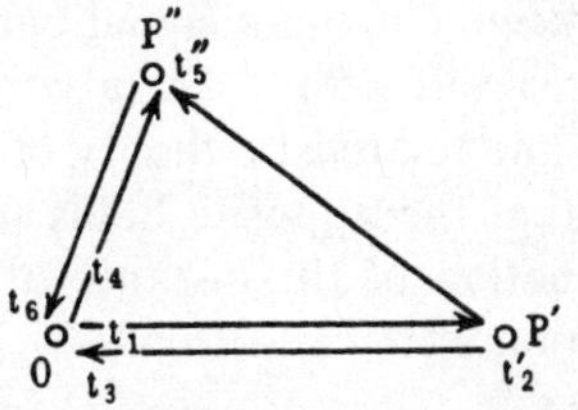

Fig. 3. Pertaining to theorem 6.

The signals $t_1 t_2' t_5''$ and $t_4 t_5''$ are two signals arriving simultaneously at P''; since the latter is a first signal, $t_1 \leqq t_4$ according to theorem 4. On the other hand, $t_2' t_5'' t_6$ and $t_2' t_3$ are two signals departing simultaneously from P'; since the latter is a first signal, $t_3 \leqq t_6$. Therefore,

$$t_2' \leqq t_5'' . \qquad (1)$$

This inference is valid for any position of the points P', P'' at any time; thus theorem 6 is proved.

---

[12] The definition by means of the mere inequality $t_1 < t_2 < t_3$ would not suffice for the proof of theorem 6.

Theorem 6 can be proved without the help of axioms II,3 and II,4, namely, as a consequence of axioms I and II,2 alone. The definition (2) of simultaneity will then have to be changed as follows. In cases such as figure 2, the value t_1 must be replaced by t, the time of departure of the last signal (last-first signal) from P which reaches P' at the time t'. The hands of the clock at P' jump at t'. This fact does not contradict theorem 6. But if one wishes to strengthen theorem 6 so that it asserts that one can give continuous values to all temporal sequences, then axioms II,3 and II,4 must be used.†

In the special instance that (1) is an equality, the time difference of the signal $\overline{P'P''} = 0$. In metrical language, the velocity of this signal becomes infinite. We can indicate what will happen in this case: both $t_1 t_2' t_5''$ and $t_2' t_5'' t_6$ must be first signals. It is also necessary that $t_2' t_5''$ be a first signal; in any case, the infinite velocity is reserved for the first signals. We could, therefore, eliminate the instance of equality in (1) by introducing the axiom that a first signal is not a signal but only a limit (cf. theorem 3). But this axiom would be false; on the contrary, it is characteristic of the relativistic theory of time in contrast to the absolute one that the opposite holds (cf. §§ 9 and 25). We obtain the elimination of the instance of equality in two steps by means of axioms II,5 and II,6.

Axiom II,5. *If $PP'P$ is a signal leaving P at the time t, there exists a $t_1 > t$ such that every signal $PP'P$ leaving at t will return to P later than t_1.*

This axiom is logically equivalent to the following theorem.

THEOREM 7. For first signals the time $\overline{PP'P} > 0$; there are no temporally closed world lines consisting of first signals.

Furthermore, we can derive the following theorem.

THEOREM 8. For any time point t at P, there exists a finite time interval at P' such that neither a signal tt' nor a signal $t't$ exists for any time point t' in this time interval.

---

† See editor's reference notes, p. 203.

Proof. Let a first signal sent from P' at the time t_1' arrive at P at the time t and return to P' at the time t_2'. According to theorem 7, $t_2' - t_1' > 0$. Consider any t_r' such that $t_1' < t_r' < t_2'$. There cannot exist a signal tt_r' because tt_2' is a first signal. Neither can there exist a signal $t_r't$, because if there were, $t_1't$ would not be a first signal. The time interval $t_2' - t_1'$ is the interval of theorem 8.

It follows from theorem 7 and the condition on ε in definition 2 for synchronizing signals OP and PO that it will always be the case that $t_1 < t_2 < t_3$. The time difference for synchronizing signals can therefore never be 0. In order to exclude this possibility also for signals $P'P''$, as in figure 3, we add the following axiom.

Axiom II,6. If for first signals, all of which leave P at the time t, $\overline{PP'P} = \overline{PP''P} = \overline{PP'P''P}$, *then* $\overline{PP'P''P} = \overline{PP''P'P}$ *for other first signal chains departing at t.*

It follows from this axiom that the addition "or at least not earlier" in theorem 6 is to be eliminated.

Proof. The condition under which the equality sign holds in (1) is $t_1 = t_4$ and $t_3 = t_6$. In that case, $t_1t_2't_3$, $t_4t_5''t_6$, $t_1t_2't_5''t_6$ are signal chains according to axiom II,6. If a first signal chain is now sent from P'' at the time t_5'' via P' to O, it will arrive at O, according to axiom II,6, at the time $t_6 = t_3$, since it may be preceded by t_4t_5''. This chain will therefore arrive (theorem 4) at P' at the time t_2'. Then $t_2't_5''t_2'$ is a temporally closed chain consisting of first signals and this result contradicts theorem 7. It is therefore impossible both that $t_1 = t_4$ and that $t_3 = t_6$; this assumption would lead to a contradiction.

We can also represent the proof by means of the following inference schema, if we employ the symbol $\supset$ (implies):

$$[t_1 = t_4, t_3 = t_6] \supset [\overline{PP'P} = \overline{PP''P} = \overline{PP'P''P}].$$

From axioms I–II,5 it follows that

$$[\overline{PP'P} = \overline{PP''P} = \overline{PP'P''P}] \supset [\overline{PP''P'P} \neq \overline{PP'P''P}],$$

otherwise, $\overline{P'P''P'} = 0$, in contradiction to theorem 7. Axiom II,6, however, says

$$[\overline{PP'P} = \overline{PP''P} = \overline{PP'P''P}] \supset [\overline{PP''P'P} = \overline{PP'P''P}].$$

Therefore,

$$[\overline{PP'P} = \overline{PP''P} = \overline{PP'P''P}] \text{ is impossible}$$

and, therefore,

$$[t_1 = t_4, t_3 = t_6] \text{ is also impossible.}$$

Axiom II,6 is a sufficient condition for the exclusion of the equality sign in (1); it is a necessary condition only if the ε in definition 2 is regarded as a constant. It does not seem to be impossible that the equals sign can be excluded by means of axiom II,5 alone, without the help of axiom II,6, if ε is varied in a suitable manner at the various space points. Axiom II,6 has a very formal character; it should be noted, however, that it holds also in absolute signal time (cf. § 25). It holds, furthermore, when P' and P'' coincide. It is also a consequence of axiom IV,2. (Nevertheless, axiom II,6 must be introduced as an axiom because axiom IV,2 holds only in special cases, whereas axiom II,6 is valid for arbitrarily moving points.) An intuitive understanding of axiom II,6 is provided by theorem 8*a* corresponding to the last but one line of the inference schema above.

THEOREM 8*a*. If $PP'P''P$ is a signal chain, there either exists a simultaneously departing signal $PP'P$ or a signal $PP''P$ (or both) which will return to P earlier than any other simultaneously departing signal $PP'P''P$.

This theorem, which presupposes axioms II,5 and II,6, can be used as an axiom in place of axioms II,5 and II,6. Axiom II,5 and theorem 9 can be derived from it, but axiom II,6 does not follow. We prefer the split into two axioms even though axiom II,6 asserts a little more.

We summarize our result in the theorem which differs from theorem 6 by the exclusion of the time interval 0 and which presupposes all of axioms I and II.

THEOREM 9. Simultaneity can be defined in such a way that a continuous positive time interval is assigned to every physical process (signal, causal chain).

Axioms I and II constitute the *topological axioms of the order of time*. Although we characterized time order by means of numerical indices in our exposition of these problems, our axioms and theorems pertain merely to topological qualities, because the applied metric is arbitrary and need not satisfy any condition except that of being compatible with the order of temporal sequences (theorem 2). Even simultaneity is at first only a topological concept so long as it is not subjected to metrical requirements (for instance, that the velocity of light be constant), and theorem 9 contains the most important *topological problem of simultaneity*.[13] Only this theorem permits the consistent application of definition 1 in a uniform continuous order of time.

The significance of the concept of simultaneity lies in a topological determination. We associate a peculiar feeling with the word "simultaneous." The idea that a distant event happens *now* has an intuitive meaning. When we analyze this feeling, we notice that it points to a singularity of the causal relation: if a distant event happens now, then neither can I influence it causally, nor can any action there have any effect on what is happening here at the present moment. This is the meaning of the intuitive idea of simultaneity. We notice, furthermore, that

[13] In this respect Lewin's presentation (see footnote to p. 27) rests on a mistake; he regards simultaneity as a necessarily metrical determination. This is the reason he misrepresents my earlier investigation, which also begins with the topological problem of time; in this publication axioms I and II are topological axioms. The clock introduced by definition 2 has only topological qualities because of the arbitrariness of the metric. The difference between Lewin's investigations and my own lies in another direction, as was revealed in a personal discussion: whereas my axioms make assertions about all *physically possible* signals and my "there exist" means "except for technical difficulties they can be produced experimentally," Lewin restricts himself to assertions about *actual* signals and does not speak of possible ones. Therefore, Lewin's investigation cannot (and will not) go as far as mine, not even topologically. Above all, the concept of first signal cannot occur in the sense of a limit. Cf. Lewin's report about his studies in *Physikalische Berichte*.

a tacit assumption is smuggled into this conception: we believe that the singularity of the causal relation characterizes one single moment at the distant place. This mistake is one of the errors of an a prioristic mode of thinking. Whether this singularity characterizes a time point or a time interval depends on the structure of causal chains which can only be determined empirically.

This topological determination distinguishes the relativistic simultaneity from the absolute one. The difference is expressed in axiom II,5 of our system; all other axioms, I to II,4, as well as II,6, hold in absolute time. Common to both conceptions of time is the validity of theorem 9, and for this reason both represent a consistent order of time. The addition "or at least not earlier" in theorem 6, though not excluded by axiom II,5 in absolute time, is excluded by another axiom (§ 25). Yet the interval $t'_2 - t'_1$ in theorem 8 derived from axiom II,5 shrinks to a point in absolute time; this theorem characterizes a single time point as simultaneous. Since this feature is absent in relativistic time, there remains a certain arbitrariness in the definition of simultaneity even if a continuous positive direction of time is retained for all causal chains. It is expressed in the arbitrariness of the factor ε in definition 2; for the absolute signal time this factor drops out because the difference $t_3 - t_1 = 0$.

§ 8. *Spatial Neighborhood and the Comparison of Neighboring Clocks*†

It is not the purpose of this investigation to discuss the topological properties of three-dimensional space, because they have been exhaustively described by mathematicians. We shall summarize all these properties in axiom V. Only "spatial neighborhood" will be defined separately.

Definition 3. A point P' is said to be in the spatial neighborhood of P with the exactness η if the time $\overline{PP'P}$ is smaller than an arbitrarily given small magnitude η.

† See editor's reference notes, p. 203.

This definition of "neighborhood" depends first of all on the temporal metric at P; for the limit $\eta = 0$, however, it becomes independent of it. Furthermore, only the "neighborhood with the exactness η" is defined; this exactness can be increased arbitrarily according to axiom I, 2. This axiom says that there are arbitrarily close neighboring points. We shall now prove theorem 10.

THEOREM 10. If P' is in the neighborhood of P, then P is in the neighborhood of P' to the same degree of exactness.

Proof. A signal is sent repeatedly back and forth between P and P'; its times are, respectively, $t_1 t_2' t_3 t_4' t_5$. According to definition 2, $(t_4' - t_2') = (t_3 - t_1) + \varepsilon(t_5 - t_3) - \varepsilon(t_3 - t_1) = (t_5 - t_3) + (1 - \varepsilon)(t_3 - t_1) - (1 - \varepsilon)(t_5 - t_3)$. If $(t_5 - t_3) = (t_3 - t_1)$, then $(t_4' - t_2') = (t_3 - t_1)$. If $(t_5 - t_3) < (t_3 - t_1)$, then in the first expression for $t_4' - t_2'$ the difference between the last two terms is negative, that is, $(t_4' - t_2') < (t_3 - t_1)$. If $(t_5 - t_3) > (t_3 - t_1)$, then in the second expression (since $(1 - \varepsilon) > 0$) the difference between the last two terms is negative, that is, $(t_4' - t_2') < (t_5 - t_3)$. According to definition 3, $(t_3 - t_1)$ as well as $(t_5 - t_3) < \eta$; therefore, in all of the three possible cases $(t_4' - t_2') < \eta$.

We shall now define the comparison of neighboring clocks.

Definition 4. Let $t = t'$ for neighboring clocks (t is the departure of the first signal from P, t' the time of its arrival at P').

The resulting error with respect to definition 2 is $<\eta$. The comparison of neighboring events is therefore independent of the temporal metric in the limit $\eta = 0$. If the synchronization at P is made from P', then according to theorem 10, the error will also be $<\eta$.

The synchronization of neighboring events is therefore symmetrical, that is, the synchronization of an event A with a neighboring event B has the same degree of exactness as the synchronization of an event B with a neighboring event A. However, this synchronization is not transitive, that is, if the event A is synchronous with a neighboring event B, and B is synchronous with a neighboring event C, both to the exactness η, then A is usually not synchronous with C to the exactness η. In this case, the error may be greater. Therefore, the synchronization of distant events cannot be replaced by a successive comparison of a number of neighboring clocks; for distant clocks the error would attain the same order of magnitude as if one set $t_2 = t_1$ instead of definition 2.

Definitions 3 and 4 do not presuppose anything about the relative state of motion of P and P'. Neighborhood and comparison of neighboring events refer to points moving arbitrarily relative to one another.

§ 9. *Fermat's Axiom*

The previous axioms state properties of arbitrary signals; they do not contain assertions concerning the special properties of light. The special character of light in the theory of relativity is stated in the following axiom.

Axiom III. First signals are direct light signals (*Fermat's axiom*).

By a "direct signal" is meant a signal without intermediate reflections; this includes the requirement that the sender at P (light source or mirror) be fixed at P, that is, that it be in the same state of motion as P. (This is defined in the same sense as "real point" above.)

Axiom II, 5 is derivable from axiom III in combination with axiom I, 1. In the complete axiom system axiom II, 5 should be canceled. We have mentioned it separately, nevertheless, in order to make axioms I and II complete in themselves. Axiom II, 4 is not derivable, because it is not definite whether or not a signal can return to P without intermediate reflection.[14] This result will follow from definition 9 and axiom V; but axiom V holds only in special coordinate systems and therefore cannot

[14] In gravitational fields the spatially closed direct light signal is possible.

justify axiom II,4 with the required generality. From axiom III the following theorem is derivable.

THEOREM 11. The time of arrival of the direct signal PP' is independent of the state of motion of P.

Proof. Let t be the time of departure of the direct signal P and t' its time of arrival at P'. If P'' and P are neighboring points, the chain of direct signals $PP''P'$ will arrive at P at a time $t_1' > t'$, since it does not represent a direct signal PP'. Let t'' be the time of arrival of this signal chain at P''; we shall now send a chain $t''PP'$ which, for the same reason, will arrive at P' at a time $t_2' > t_1'$. Let t_1 be the time of arrival of this chain at P. According to definition 3, $t_1 - t < \eta$. If η converges toward 0, then $t_2' - t'$ likewise converges toward 0 according to theorem 4. Since $t_2' > t_1' > t'$, the times of arrival of the two first signals PP' and $P''P'$ coincide in the limit for infinitesimal neighborhood relations. Yet the first signal $P''P'$ corresponds to a state of motion of the emitting point different from that to which PP' corresponds, because $P''P'$ is reflected by a mirror set up at P'' or has been emitted by a light source fixed at P''. Since P'' and P are in arbitrary states of motion relative to each other, the state of motion of the emitting point is irrelevant for the time of arrival of the first signal.

Only in combination with axiom III do axioms I and II constitute a description of the motion of light whose topological qualities they formulate. Theorems 4 and 11 express an essential topological law of the motion of light which we shall call the *continuity of the transmission of light*. Axiom III, in combination with axiom II,5 (derivable from axiom III), formulates the second important topological property, the *limiting character of the propagation of light*.

§ 10. *Axiom of Stationarity*

It will now be necessary to make a selection among the real points, that is, to combine certain real points into a *spatial coordinate system.*[15] Rigid connections are usually introduced for this purpose. It can be shown, however, that the lines between two points can be defined by means of light signals alone; axioms IV and V contain the properties of light necessary for this purpose. They refer to certain selected point systems. Since they introduce a metric into the order of time, they are also called *metrical light axioms.*

Definition 5. A system of real points is called "everywhere present" if there does not exist a real point which does not at every time point (determined on it) neighbor a real point of the system with an arbitrary exactness η.

Definition 6. Let an arbitrary metric be chosen at A as a measure of equal time intervals. Let us seek all those (arbitrarily moving) real points P for which (according to this metric) the time $\overline{APA}$ *of first signals does not change with time. This point system is called "a system related to A."*

Axiom IV,1. It is possible to choose the metric at A in such a way that the resulting point system related to A is, for each of its points, a related system. Such a point system is everywhere present.

Definition 7. A point system, conforming to axiom IV,1, is called a stationary spatial system;[16] *a clock synchronous with it has stationary time.*

Simultaneity is defined according to definition 2 with $\varepsilon = \frac{1}{2}$.

Definition 8. Let a first signal be sent from A at the time t_1; *reflected at B, let it return to A at the time* t_3. *Then its time of arrival at B will have the following value:*

$$t_2 = \frac{t_1 + t_3}{2}$$

(*Einstein's definition of simultaneity*).

[15] This is the reason we no longer call the selected real points $P, P' \ldots$, but $A, B, C \ldots$.

[16] For the time being we are defining only the stationary coordinate system; the general coordinate system will be defined in definition 23.

The advantage of setting $\varepsilon = \frac{1}{2}$ consists in the fact that with axiom IV,1 the synchronism becomes symmetrical and with axiom IV,2 transitive.[17]

Proof of symmetry. "Symmetry of synchronism" means that if B is synchronous with A, then A is synchronous with B. For this proof we set the clock A from B. A light signal travels back and forth several times; its times at A are $t_1 t_3 t_5$ and at B, correspondingly, $t_2' t_4'$. According to definition 8,

$$t_2' = \frac{t_1 + t_3}{2} \qquad t_4' = \frac{t_3 + t_5}{2}.$$

According to axiom IV, 1, however, for every B

$$t_3 - t_1 = t_5 - t_3.$$

There are three equations for the five variables $t_1 \ldots t_5$; if t_1 and t_5 are eliminated,

$$t_3 = \frac{t_2' + t_4'}{2}$$

remains as the relation between $t_2' t_3 t_4'$. This means that clock A simply shows the times according to which it would have to be set if it were to be made synchronous with B. Since B was synchronized with A, A is also synchronous with B.

If at the time t_5 the signal is sent once more on a round trip, the following theorem can be derived.

THEOREM 12. $\overline{ABA} = \overline{BAB} =$ *constant.*

If a clock were placed at A whose rate is a function of higher than first degree of the first clock, then $\overline{ABA}$ would no longer be constant for all points B of the system. Since A determines the rate of the other clocks, the rate of the clocks is determined up to a linear function. Only the unit and the zero point are

---

[17] In his first paper on the theory of relativity (*Ann. d. Phys.*, vol. 17 [1905], p. 891) A. Einstein treated definition 8 as a definition; he noted, however, that the symmetry and transitivity of the resulting synchronism is not a matter of definition but a matter of fact. He is therefore the originator of this important distinction. In subsequent lectures he made use of the round-trip axiom VI, 2 for the justification of transitivity, and thus he is also the author of this axiom.

arbitrary and a period is characterized as uniform. This period depends on the relative states of motion of the points *A* and *B* and since the stationary system is not necessarily rigid, the resulting rate is not yet identical with that of natural clocks. This result will be reached when the stationary system is also subjected to the conditions of axioms IV,2 and V; in this case, uniformity is defined by the motion of light alone. The aim that classical mechanics reached only inadequately by means of the law of inertia—inadequately because by means of the force-free translation the beginning and end of the period are measured at spatially distant points, that is, by means of a method presupposing the definition of simultaneity—is reached much more simply by the motion of light. The definition of simultaneity is not contained in axioms I to V.

Definition 8 determines simultaneity at distant points so that it is dependent upon the choice of the rate of the central clock. Since in a stationary system the rate of a clock is prescribed up to a linear function, simultaneity for these systems is uniquely defined. It is easily calculated that two simultaneous events remain simultaneous even if the rate of the clocks changes by a linear function.

The *time interval* between two points[18] *A* and *B* is the time difference between the departure of a signal at *A* and its arrival at *B*. If the time interval is measured by means of clocks synchronized according to definition 8, then this interval is constant because of axiom IV,1 and equal in both directions. If clocks are synchronized along a polygon path *A*, *B*, *C*...*H*, the time interval is constant for each partial distance *AB*, *BC*, *CD*.... Therefore, the sum of the times $\overline{ABC...H...CBA}$ is constant. This sum can be measured at *A*; the clocks at *B*, *C*...*H* are not relevant. Therefore, the assertion

$$\overline{ABC...H...CBA} = constant$$

is valid independently of the definition of the synchronism.

[18] It is to be noted that the time interval is defined without the concept of distance.

This conclusion is equally valid if H is made to coincide again with A. Imagine a second clock U' established at A and synchronized with the first clock U, not directly through a neighborhood comparison, but via the polygon path $ABC\ldots A\ldots CBA$. Then $\overline{ABC\ldots A\ldots CBA} = constant$, measured by U. For this final result U' is irrelevant. Now the question arises: Is $\overline{ABC\ldots A} = constant$, too?

Let us call the round-trip time along the polygon path a in one direction and b in the other direction; let us distinguish the measurement of this time by means of U' from the measurement by means of U by adding a prime to the symbols. In general, $a \neq b$; the question is whether all a's, compared to one another, and all b's, compared to one another, are equal. Let the signals $a_1 b_2 a_3$ depart one after the other. From the conclusion concerning polygon paths and from theorem 12 the following equation is derivable:

$$(a_1 + b_2) = (b_2' + a_3') = constant. \qquad (2)$$

On the other hand, $b_2 + a_3 = constant$, too, because the polygon path can be traveled in the opposite direction from U.

$$\begin{array}{lll}
\kappa & a_1 & \\
 & b_2 & \lambda \\
\kappa & a_3 & \\
 & b_4 & \lambda \\
 & a_5 & \\
 & \vdots & \\
 & b_{r-2} & \lambda \\
\kappa & a_{r-1} & \\
 & b_r &
\end{array}$$

This does not mean necessarily that $b_2' + a_3' = b_2 + a_3$; but since the measurement of this identical path furnishes a constant at U as well as at U', then U and U', seen in a neighborhood comparison, can differ only by a linear function. We now send signals back and forth along the polygon path, according to the foregoing schema. All times are measured by U; thus

$$\kappa = a_i + b_{i+1} = constant \text{ and } \lambda = b_k + a_{k+1} = constant. \quad (3)$$

If the number r (r is even) is very large, then in the total sum

$$\frac{r}{2} \cdot \kappa = \left(\frac{r}{2} - 1\right) \cdot \lambda + (a_1 + b_r)$$

the last term is small compared to the last but one term because, according to axiom I, 1, $a_1 < \lambda$ and $b_r < \lambda$. On the other hand, a_1 and b_r are positive, according to axiom I, 1. Therefore, the following inequalities hold:

$$\kappa < \left(1 - \frac{2}{r}\right)\lambda + \frac{4\lambda}{r} \qquad \kappa > \left(1 - \frac{2}{r}\right)\lambda . \qquad (4)$$

These inequalities can be satisfied for an arbitrarily large r only if $\kappa = \lambda$; that is, if $\kappa \neq \lambda$, there exists an r after which one of the inequalities will no longer hold. If $\lambda = \kappa + \delta$, $\delta > 0$, then the second equation results in the condition

$$r < \frac{2(\kappa + \delta)}{\delta},$$

that is, it no longer holds for larger r. Analogous results are obtained from the first equation for $\delta < 0$.

From (2) and (3) it follows that

$$b_2 + a_3 = b_2' + a_3' .$$

Since U and U' furnish the same coefficient of measure for this identical process, their unit must be the same. Their time indications, viewed in a neighborhood comparison, can vary only by an additive constant. If the time of departure of a signal in the direction a measured at $U = t_1$ and the time of arrival measured at $U_1' = t_2'$, then the travel time $t_2' - t_1 = constant$ for synchronized clocks. If the same time of arrival measured at $U = t_2$, then in a neighborhood comparison $t_2 = t_2' + k$, where $k = constant$. Therefore, $t_2 - t_1 = constant$ and the following theorem holds.

THEOREM 13. The simple round-trip time of a polygon path $\overline{ABC\ldots A}$ is constant.

If all clocks are synchronized directly from A, then the time interval $\overline{AB} = \overline{BA} = constant$ between A and any other point B is constant. Because of theorem 13, the time interval $\overline{BC}$ is constant, too, since $\overline{AB} + \overline{BC} + \overline{CA} = constant$. Thus we arrive at theorem 14.

THEOREM 14. The travel time between any two points is constant.

In the system constructed according to axiom IV, 1 stationary relations hold for light signals; therefore, we have called it the *axiom of stationarity*. We still do not have $\overline{BC} = \overline{CB}$; this condition, important for the principle of the constancy of the velocity of light, will be obtained by means of the following axioms.

§ 11. *The Round-Trip Axiom*

Axiom IV, 2. It is possible to select among the stationary systems a "static" one possessing the following property. If from a point A of the static system two light signals are sent in opposite directions along a closed triangular path ABCA, they will return simultaneously.

The axiom asserts the equality of the time intervals of the round trip merely for signals departing simultaneously from A. Only by means of theorem 13, that is, of axioms I to IV, 1, can the equality at all times be derived.

By means of axiom IV, 2 we can conclude that the synchronism is transitive. Let A be synchronous with B and A synchronous with C; it is to be proved that B is also synchronous with C.[19]

[19] If the synchronism were not symmetrical, the synchronism of B and C could be derived only from B being synchronous with A and A being synchronous with C.

Proof of transitivity (fig. 4). A light signal is sent in the direction $ABCA$; the second signal, traveling in the opposite direction, is not sent simultaneously but so that it will arrive at C at the same time as the first signal. The intervals $\overline{AB}$, $\overline{BC}$, and $\overline{CA}$ are $a_1 a_2 a_3$, and $b_3 b_2 b_1$ in the other direction. Because of the synchronization from A

$$a_1 = b_1 \qquad a_3 = b_3.$$

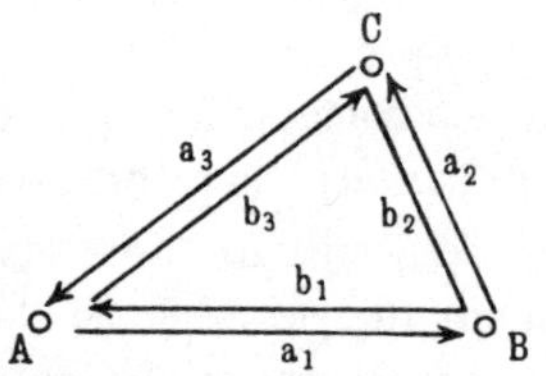

Fig. 4. Pertaining to the proof of transitivity of synchronism.

According to axiom IV,2 and theorem 13, however,

$$a_1 + a_2 + a_3 = b_1 + b_2 + b_3$$

and, therefore,

$$a_2 = b_2. \tag{5}$$

These two time intervals were chosen so that they would correspond to a signal reflected directly at C, because the signal CB leaves C the moment BC arrives. Formula (5) says that this moment at C has a time index corresponding to half the time interval between the time of departure and the time of arrival of a signal BCB; therefore, C is synchronous with B. If the time of departure of the second signal is chosen so that it will arrive at B simultaneously with the first signal, then B is synchronous with C. Therefore, transitivity holds.

Equation (5) may be formulated also as follows.

THEOREM 15. The time interval between any two points of the static system is equal in both directions and at all times.

This theorem follows directly from the symmetry and the transitivity.

§ 12. *The Space Metrical Light Axiom*

We will consider the spatial geometry of this system only after we have defined a system of real points by means of the measurement of time, that is, after we have established a spatial coordinate system in which time is defined at every place. Its metrical relations can be only defined, not discovered, and this definition is another *coordinative definition* coordinating certain physical things to geometrical concepts. Light will once more be the only physical entity used for the measurement of space. It follows from the specific facts connected with the propagation of light laid down in axioms I to V that the given definition leads to certain simplicity properties of the resulting geometry. These facts explain why the propagation of light is sufficient for a unique definition of spatial geometry.

Definition 9. Light rays are straight lines.

A small bundle of light rays emitted from *A* may indicate the direction of the ray; all points hit by the bundle without an interference of mirrors lie on a straight line. This definition is strict for the transition to the limit zero for the opening of the bundle. (It coincides with the direction of the wave normal because of the definition of simultaneity we employ; but this fact depends on axiom V.)

Axioms IV, in combination with the other axioms, state those properties of light rays which support definition 9: The light rays satisfy the requirements of geometry for straight lines. In this connection we shall prove the following two theorems. They do not assert yet that through two points there is only one straight line; this requirement holds only for Euclidean geometry and will be laid down in axiom V.

THEOREM 16. If a direct signal travels once from *A* via *C* to *B*, then a direct signal always travels from *A* via *C* to *B*.

Proof. Let the time be set at every point according to definition 8. The direct signal is sent from *A* via *C* to *B* (fig. 5). Departure at *A*: t_1, arrival at *B*: t_2. Let $\Delta t = t_2 - t_1$. Let us first assume that theorem 16 is not valid and let us pursue a signal that departs from A at a later time t_1' and, reflected at C at t_x', arrives at B at t_2'. According to theorem 14, the time intervals $\overline{AC}$ and $\overline{CB}$ are constant; therefore, $(t_x' - t_1') + (t_2' - t_x') = t_2' - t_1' = \Delta t$. If a direct signal is sent simultaneously to B, it will also arrive at B at the time t_2' since the time interval $\overline{AB}$ is constant. The signal reflected at C will therefore arrive at B simultaneously with the direct signal. It is a first signal and a direct signal according to axiom III. Therefore, theorem 16 is valid. For this proof axioms I–IV, 1 suffice.

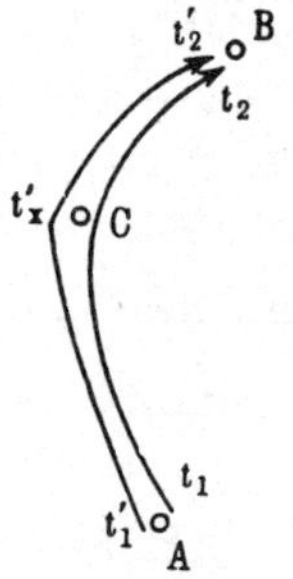

Fig. 5. Pertaining to theorem 16.

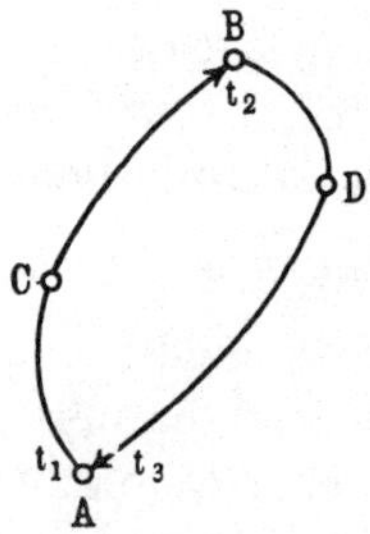

Fig. 6. Pertaining to theorem 17.

THEOREM 17. If a direct signal travels from A via C to B, then a direct signal travels also from B via C to A.

Proof. By using axiom IV,2 also, we shall presuppose the transitivity of the synchronism. At first, we shall assume theorem 17 to be invalid. We send a direct signal from A via C to B (fig. 6) and, reflected at B, return it as a direct signal via D to A; departure and arrival at A: t_1, t_3; arrival at B: t_2. According to definition 8, $t_2 = \frac{1}{2}(t_1 + t_3)$. This path is identical with the triangular path of direct signals $ABDA$; therefore $\overline{ACBDA} = \overline{ABDA}$. According to axiom IV, 2, this is $= \overline{ADBA}$, and, according to the assumption of this proof, this is identical with $\overline{ADBDA}$. Since the synchronism is transitive, $\overline{AD} = \overline{DA}$ and $\overline{DB} = \overline{BD}$. Since $\overline{ADBDA} = \overline{ACBDA}$, the signal ADB

arrives at B at the time $t_2 = \frac{1}{2}(t_1 + t_3)$. Thus it arrives at B simultaneously with the direct signal ACB; it is therefore a first signal and according to axiom III a direct signal. Thus we have shown that a direct signal travels back and forth along the path ADB. Analogous inferences can be made for the path BCA. $\overline{ACBDA} = \overline{ACBA} = \overline{ABCA} = \overline{ACBCA}$; therefore, the signal BCA departs from B simultaneously with the direct signal BDA and arrives at A simultaneously with this signal. Therefore, according to axiom III, BCA is also a direct signal.

We shall now define *equality of interval* and *length of interval.*

Definition 10. If two signals depart simultaneously from A along the paths ABA and ACA and return simultaneously to A, then the interval $AB = AC$.

Definition 11. The length of an interval AB is measured by $\frac{c}{2}\overline{ABA}$. c is an arbitrary constant.

Because of the validity of axiom III these two definitions are compatible with definition 9 of *straight line*, that is, they satisfy the conditions of the geometry for the concept of length; without this axiom the straight line would not be the shortest connection between two points. For the light geometry, Fermat's principle is therefore essential.

With the definition of the measurement of length, velocity is now defined as the quotient of distance and transit time. From axioms I–IV and the given definitions (definitions 8, 9, 10, and 11 are relevant) together with theorem 15, the following theorem can be derived.

THEOREM 18. The velocity of light within the given coordinate system is constant and $= c$.

Only now shall we introduce an assumption concerning the resulting geometry.

Axiom V. It is possible to choose the static system relative to A so that the defined spatial geometry will become three-dimensionally Euclidean.

Definition 12. A stationary spatial coordinate system conforming to axioms I to V is called an inertial system; the points of such a system are called at rest relative to one another and the corresponding stationary time is called uniform.

The requirement that the geometry be Euclidean results in a selection among the systems still possible according to axioms IV, 1 and IV, 2. We shall show in the following paragraphs that this additional requirement is sufficient to prescribe uniquely the state of motion of the points of the system relative to one another; thus the definition of the points at rest relative to one another and of uniform time is unique.

In our arrangement of the definitions the spatial geometry is not arbitrary but an empirical fact, because equality of segment was defined independently of the geometry. At the same time, this result illustrates the fact that Euclidean geometry is not presupposed in every comparison of congruence. Congruence was defined in a different way, without the use of rigid bodies—by means of the velocity of light alone.

§ 13. *Proof of the Uniqueness of Definition* 12 *of the Relative Rest of Real Points*

We shall present the proof in the following manner. We assume a given system K for which axioms I to V hold. At A we introduce an arbitrary function $\tau = f(t)$ of time and form the point system K' relative to A originating with this new time. We assume that in K' all axioms I to V hold; if we can show that the function $\tau = f(t)$ must then be linear, we have proved the uniqueness of rigidity.

It is sufficient to prove this theorem for spatially two-dimensional systems K and K', because then it will hold, a fortiori, for three-dimensional systems.[20] We introduce polar coordinates in K: r, φ, t, in K': ρ, ϑ, τ, and search for the transformation formulas from K into K'. We may immediately

[20] However, it is not valid for one-dimensional systems; here the condition of Euclidean geometry fails.

set

$$\vartheta = \varphi \tag{6}$$

because if K' revolved relative to K, theorem 17 would not hold in K'.[21] All points of K' must move radially from A or toward A. In order to find the transformations, we shall follow the path of a key signal from A to B' and back (B' at rest in K'); its times in K are (with the velocity of light $c = 1$): $t - r$, t, $t + r$. According to the τ clock established at A, the times of departure and return are $f(t - r)$ and $f(t + r)$. According to definition 8, in K' the arrival at B will have the time

$$\tau = \tfrac{1}{2}[f(t + r) + f(t - r)]. \tag{7}$$

This is the time transformation. If the velocity of light in K' is also set $c = 1$, the distance AB' becomes

$$\rho = \tfrac{1}{2}[f(t + r) - f(t - r)]. \tag{8}$$

Formulas 6 to 8 represent the desired transformations.

We shall now formulate the condition for the validity of axiom IV, 1 in K'. We send a signal from B' to C' lying on another radius; the time difference between departure from B' and arrival at C', measured by clocks established in K', becomes

$$\tau_2 - \tau_1$$
$$= \tfrac{1}{2}[f(t_2 + r_2) + f(t_2 - r_2) - f(t_1 + r_1) - f(t_1 - r_1)], \tag{9}$$

if $t_1 r_1$ and $t_2 r_2$ are the coordinates of these point events in K. If one follows the path of the signal in K and if $\alpha = cos\ \varphi$, then the following relation holds between the coordinates:

$$(t_2 - t_1)^2 = r_1^2 + r_2^2 - 2r_1 r_2 \alpha. \tag{10}$$

Furthermore, the radial coordinates of B' and C' in K' are

$$\rho_1 = \tfrac{1}{2}[f(t_1 + r_1) - f(t_1 - r_1)], \tag{11}$$

$$\rho_2 = \tfrac{1}{2}[f(t_2 + r_2) - f(t_2 - r_2)]. \tag{12}$$

---

[21] This fact is due to the failure of the round-trip axiom IV, 2.

If axiom IV,1 is to hold in K', then, according to theorem 14, $\tau_2 - \tau_1 =$ *constant* if $\rho_1\rho_2\alpha$ are constant; that is, $\tau_2 - \tau_1 =$ *constant* for every value system $t_1r_1t_2r_2$ which satisfies conditions (10) to (12). This is a condition for the function f; if three of the magnitudes $t_1r_1t_2r_2$ are assumed to be eliminated from (9) by means of the three equations (10) to (12), then (9) = *constant* is a functional equation for f.

If we assume the validity of axioms IV,2 and V in K', we can add another condition. Then the cosine theorem of the Euclidean plane triangle must hold in K', thus corresponding to (10):

$$(\tau_2 - \tau_1)^2 = \rho_1^2 + \rho_2^2 - 2\rho_1\rho_2\alpha. \tag{13}$$

Then (9) $\equiv \sqrt{\rho_1^2 + \rho_2^2 - 2\rho_1\rho_2\alpha}$ is the functional equation for f.

In order to make the mathematical part of this consideration clearer, we introduce the substitutions

$$\left.\begin{aligned} \rho_1 &= \frac{\eta_2}{2} & t_1 - r_1 &= t \\ \rho_2 &= \frac{\eta_3 - \eta_1}{2} & t_2 - r_2 &= t + \delta_1 \\ & & t_1 + r_1 &= t + \delta_2 \\ \tau_2 - \tau_1 &= \frac{\eta_1 + \eta_3 - \eta_2}{2} & t_2 + r_2 &= t + \delta_3. \end{aligned}\right\} \tag{14}$$

If ρ_1, ρ_2, $\tau_2 - \tau_1$ are constant, then $\eta_1\eta_2\eta_3$ are constant, too.

If we substitute these magnitudes in (9), (11), and (12) and add the definition $\tau = f(t)$, we obtain

$$\begin{aligned} f(t) &= \tau, \\ f(t + \delta_1) &= \tau + \eta_1, \\ f(t + \delta_2) &= \tau + \eta_2, \\ f(t + \delta_3) &= \tau + \eta_3. \end{aligned}$$

If the inverse function to f is called F and if the substitutions (14) are introduced also in (10) and (13), we obtain the following system of equations for F:

$$\left.\begin{aligned} F(\tau) &= t \\ F(\tau + \eta_1) &= t + \delta_1 \\ F(\tau + \eta_2) &= t + \delta_2 \\ F(\tau + \eta_3) &= t + \delta_3 \end{aligned}\right\} \tag{15}$$

$$\alpha = \frac{\delta_1\delta_2 + \delta_2\delta_3 - 2\delta_1\delta_3}{\delta_2(\delta_3 - \delta_1)}, \tag{16}$$

$$\alpha = \frac{\eta_1\eta_2 + \eta_2\eta_3 - 2\eta_1\eta_3}{\eta_2(\eta_3 - \eta_1)}. \tag{17}$$

This means that if the three magnitudes $\eta_1\eta_2\eta_3$ are given arbitrarily, they determine three magnitudes $\delta_1\delta_2\delta_3$ by means of (15) and a magnitude α by means of (17). F is to have the

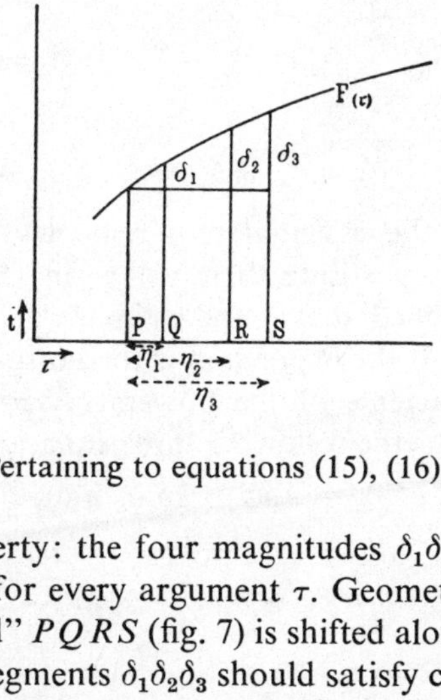

Fig. 7. Pertaining to equations (15), (16), and (17).

following property: the four magnitudes $\delta_1\delta_2\delta_3\alpha$ must satisfy condition (16) for every argument τ. Geometrically speaking, if the rigid "grid" $PQRS$ (fig. 7) is shifted along the horizontal axis, then the segments $\delta_1\delta_2\delta_3$ should satisfy continuously condition (16) whose constant α is given by the analogous construction (17) of the η_i. It can be shown that this requirement demands the linearity of F.

Proof. We shall simplify the proof by setting $\alpha = 0$, that is, $\varphi = 90°$. If the linearity of F can be derived for this particular position of the radii ρ_1 and ρ_2, it would certainly hold in the general case.

We shall give the proof by obtaining conditions for the derivations from F. We shift the grid by $d\tau$; if we set $\tau + \eta_1 = \tau_1$, $\tau = \eta_2 = \tau_2$, $\tau + \eta_3 = \tau_3$, then

$$\begin{aligned} F(\tau + d\tau) &= t + dt_0 & dt_0 &= F'_0\, d\tau \\ F(\tau_1 + d\tau) &= t + \delta_1 + dt_1 & dt_1 &= F'_1\, d\tau \\ F(\tau_2 + d\tau) &= t + \delta_2 + dt_2 & dt_2 &= F'_2\, d\tau \\ F(\tau_3 + d\tau) &= t + \delta_3 + dt_3 & dt_3 &= F'_3\, d\tau \end{aligned} \qquad F' = \frac{dF}{d\tau}.$$

If the grid is shifted, new magnitudes δ'_i replace the δ_i; they become

$$\delta'_1 = \delta_1 + dt_1 - dt_0 = \delta_1 + (F'_1 - F'_0)\, d\tau.$$

Corresponding magnitudes result for δ'_2 and δ'_3. If the β_i are defined by

$$\beta_1 = F'_1 - F'_0 \qquad \beta_2 = F'_2 - F'_0 \qquad \beta_3 = F'_3 - F'_0, \tag{18}$$

then

$$\delta'_1 = \delta_1 + \beta_1\, d\tau \qquad \delta'_2 = \delta_2 + \beta_2\, d\tau \qquad \delta'_3 = \delta_3 + \beta_3\, d\tau. \tag{19}$$

The δ'_i—this is the essential requirement—must once more satisfy (16); we substitute their values in (16) while setting $\alpha = 0$, as mentioned above, and establish an order according to powers of $d\tau$. If the expression is to vanish for every choice of $d\tau$, the coefficients of the powers of $d\tau$ must disappear individually. Thus the following three equations result:†

$$\delta_1\delta_2 + \delta_2\delta_3 - 2\delta_1\delta_3 = 0, \tag{20}$$

$$\beta_1\beta_2 + \beta_2\beta_3 - 2\beta_1\beta_3 = 0, \tag{21}$$

$$\delta_1\beta_2 + \delta_2\beta_1 + \delta_3\beta_2 + \delta_2\beta_3 - 2\delta_3\beta_1 - 2\delta_1\beta_3 = 0. \tag{22}$$

---

† Cf. Hans Reichenbach, "Über die physikalischen Konsequenzen der relativistischen Axiomatik," in *Zeitschr. f. Physik*, vol. 34 (1925), p. 34, which offers the following correction (translation mine):

"In this connection, a mistake occurs in *AR* which, even though not essential for the construction as a whole, I should like to correct on this occasion. Although the distinction of the classes S and J is made correctly, more is derived from the light geometry than is actually warranted because of a mathematical error. Let us symbolize the class of the system S not belonging to J by 'T'; written in symbols this becomes $J + T = S$.

If $\tau = 0$ (the choice of the zero point is irrelevant), the following relation derived from (17) can be added:

$$\tau_1\tau_2 + \tau_2\tau_3 - 2\tau_1\tau_3 = 0. \tag{23}$$

Equations (21) and (22) are conditions for the derivatives β_i. We can obtain further conditions for the β_i from (20) and (23). Let us consider another variation of the grid. Since the three magnitudes $\eta_1\eta_2\eta_3$ were arbitrary, each one of the points Q, R, S can be shifted certain distances $d\tau_1 d\tau_2 d\tau_3$. This variation can be arranged in such a way that the condition $\alpha = 0$ remains satisfied for the η_i; for the $d\tau_1 d\tau_2 d\tau_3$, this means a dependence resulting from a variation of (23):

$$(\tau_2 - 2\tau_3)\, d\tau_1 + (\tau_1 + \tau_3)\, d\tau_2 + (\tau_2 - 2\tau_1)\, d\tau_2 = 0. \tag{24}$$

According to this variation, the δ_i must once more satisfy (16); therefore, the changes $d\delta_i$ must satisfy the condition resulting from the variation of (16). Now $d\delta_i = dt_i = F_i' d\tau_i$; if $F_0' = k$, then by means of (18) $d\delta_i = (\beta_i + k)d\tau_i$. By means of a variation of (16) we obtain

$$(\delta_2 - 2\delta_3)(\beta_1 + k)\, d\tau_1 + (\delta_1 + \delta_3)(\beta_2 + k)\, d\tau_2 + (\delta_2 - 2\delta_1)(\beta_3 + k)\, d\tau_3 = 0. \tag{25}$$

The following view is maintained in *AR*: a system *J* and a system *T* cannot permanently share a spatial point. This statement is false.

"It leads to the following thesis. If an arbitrary point *A* is chosen as starting point and if a system *S* is constructed around it according to the rules of the light geometry, this system is *uniquely* determined; either a system *J* or a system *T* can be obtained, depending on the state of motion of *A*. In this form I proved the theorem, but this proof contains an error, the discovery of which I owe to John von Neumann, Zürich. Equation (21) on p. 58 in *AR* is false, because in it the coefficient of the term of second order is evaluated whereas (19) is valid merely in first order. The proof is therefore invalid. The second proof of the theorem, given by Mr. K. Fladt, also contains a mistake (*AR*, p. 62); the form following (31) is false. Here, numerator and denominator of the fraction on the right individually need not be independent of η_1. Because of this error, the general solution of the theorem of figure 7 in *AR* is not the linear function $A\tau + B$, but the fractional rational function $A\tau + B/C\tau + D$. Von Neumann proved it by correcting the mistake through an adequate extension of Fladt's proof. This is the reason why already at this place the viewpoint of my presentation in *AR*, § 16, considering the systems *T*, must be taken into account. In contradiction to the assertion in *AR*, p. 81, more than one system in which light axioms hold can be constructed around a fixed starting point."—*Ed.*

(24) cannot represent a new condition for the $d\tau_i$ but is to be satisfied for any choice of the $d\tau_i$ compatible with (23); therefore, (24) $\equiv$ (25). Thus the coefficients of (25) must result from the coefficients of (24) through multiplication by a factor $\lambda = \lambda(\tau_1\tau_2\tau_3)$; therefore,

$$\left.\begin{aligned} \beta_1 &= \lambda \cdot \frac{\tau_2 - 2\tau_3}{\delta_2 - 2\delta_3} - k \\ \beta_2 &= \lambda \cdot \frac{\tau_1 + \tau_3}{\delta_1 + \delta_3} - k \\ \beta_3 &= \lambda \cdot \frac{\tau_2 - 2\tau_1}{\delta_2 - 2\delta_1} - k. \end{aligned}\right\} \qquad (26a)$$

Here τ_3 is eliminated by means of (23) and δ_3 by means of (20). The four remaining magnitudes $\tau_1\tau_2\delta_1\delta_2$ can be reduced to three magnitudes γ, a, b, since they occur in quotients:

$$\frac{\tau_1}{\delta_1} = \gamma, \qquad \tau_2 = b \cdot \tau_1, \qquad \delta_2 = a \cdot \delta_1. \qquad (27)$$

From (26a) are derivable

$$\left.\begin{aligned} \beta_1 &= \lambda \cdot \left(\frac{\tau_2}{\delta_2}\right)^2 \cdot \frac{\delta_2 - 2\delta_1}{\tau_2 - 2\tau_1} - k = \lambda \cdot \gamma \cdot \frac{b^2}{a^2} \cdot \frac{a - 2}{b - 2} - k \\ \beta_2 &= \lambda \cdot \left(\frac{\tau_1}{\delta_1}\right)^2 \cdot \frac{\delta_2 - 2\delta_1}{\tau_2 - 2\tau_1} - k = \lambda \cdot \gamma \cdot \frac{a - 2}{b - 2} - k \\ \beta_3 &= \lambda \cdot \frac{\tau_2 - 2\tau_1}{\delta_2 - 2\delta_1} - k \qquad\quad = \lambda \cdot \gamma \cdot \frac{b - 2}{a - 2} - k. \end{aligned}\right\} \qquad (26b)$$

These are three further conditions for the β_i which are added to (21) and (22). From the conjunction of these five equations we obtain a condition for F. We substitute (26b) into (21) and (22) while expressing δ_3 by δ_1 and δ_2 in (21) and (22) according to (20) and obtain

$$\left.\begin{aligned} A\lambda\gamma - Bk &= 0 \\ C\lambda\gamma - Dk &= 0, \end{aligned}\right\} \qquad (28)$$

where

$$A = \frac{b^2}{a^2} \cdot \frac{(a-2)^2}{(b-2)^2} + 1 - 2\frac{b^2}{a^2},$$

$$B = 2 \cdot \frac{a-2}{b-2} - \frac{b^2}{a^2} \cdot \frac{a-2}{b-2} - \frac{b-2}{a-2},$$

$$C = \frac{a-2}{b-2} + \frac{b^2}{b-2} - \frac{a}{b-2} + b - 2,$$

$$D = \frac{2(a^2 - 2a + 1)}{a-2}.$$

The two equations (28) are homogeneous and linear in $\lambda\gamma$ and k. Since k and $\lambda\gamma$ are not $= 0$, the determinant of their coefficients must vanish. Therefore,

$$AD - BC = 0.$$

This is a condition for a and b alone; it results from the conjunction of conditions (20) to (25). By means of elementary computation it leads to

$$(a - b)^4 = 0,$$

that is, to $a = b$. According to (27), F is a linear function, since it must be satisfied for any arbitrary choice of the arguments τ_1 and τ_2. Then f is linear, too.

This result demonstrates that definition 12 of uniform time is unique, that is, that uniform time, except for a linear function, can be prescribed by light signals alone. At the same time, the uniqueness of the defined relative rest of the points is proved.

The requirement of the validity of Euclidean geometry is essential for this proof. It is possible to specify a system that is not uniquely rigid and yet satisfies axioms I to IV; only it is not Euclidean. Such a system can be obtained for K' if axiom V is omitted in K'. Then the cosine theorem of the plane triangle is not valid in K' and the magnitude α in (17) is not identical with the α of (16). (17) is eliminated, but the other conditions (15) to (16) for F remain the same. It is seen easily that the theorem represented in figure 7 can also be satisfied by a nonlinear function, namely, by $F(\tau) = e^\tau$. If the grid is shifted, that is, if the constant k is added to the argument τ, the δ's are multiplied by e^k and as a consequence (16) remains satisfied.

K. Fladt, with whom I discussed this question, has proved that the exponential function constitutes the only nonlinear solution of this generalized theorem. I publish this proof here by his kind permission. (16), transformed, together with (15), yields

$$\frac{1+\alpha}{2} = \frac{\delta_3(\delta_2 - \delta_1)}{\delta_2(\delta_3 - \delta_1)}$$

$$= \frac{F(\tau+\eta_3) - F(\tau)}{F(\tau+\eta_2) - F(\tau)} \cdot \frac{F(\tau+\eta_2) - F(\tau+\eta_1)}{F(\tau+\eta_3) - F(\tau+\eta_1)}. \quad (29)$$

This expression is meant to be independent of τ. For instance, it should be valid for $\tau = 0$, that is, it is the following function of the η alone:

$$\frac{F(\eta_3) - F(0)}{F(\eta_2) - F(0)} \cdot \frac{F(\eta_2) - F(\eta_1)}{F(\eta_3) - F(\eta_1)}. \quad (30)$$

If the right-hand side of (29) is set equal to (30), we obtain

$$\frac{F(\tau+\eta_3) - F(\tau)}{F(\tau+\eta_2) - F(\tau)} \cdot \frac{F(\eta_2) - F(0)}{F(\eta_3) - F(0)}$$

$$= \frac{F(\tau+\eta_3) - F(\tau+\eta_1)}{F(\tau+\eta_2) - F(\tau+\eta_1)} \cdot \frac{F(\eta_2) - F(\eta_1)}{F(\eta_3) - F(\eta_1)}.$$

The expression on the right is independent of η_1, since the expression on the left is. It is therefore a function of τ, η_2, η_3 alone, i.e., $= H(\tau, \eta_2, \eta_3)$. Since it can be brought into the form

$$\frac{F(\tau+\eta_3) - F(\tau+\eta_1)}{F(\eta_3) - F(\eta_1)} : \frac{F(\tau+\eta_3) - F(\tau+\eta_1)}{F(\eta_2) - F(\eta_1)}, \quad (31)$$

then

$$H(\tau, \eta_2, \eta_3) = \frac{K(\tau, \eta_3)}{K(\tau, \eta_2)},$$

where $K(\tau, \eta_3)$ and $K(\tau, \eta_2)$ are given by the two expressions to the left and to the right of the colon in (31). If η_1 and η_3 are interchanged in the expression on the left, nothing changes; therefore, $K(\tau, \eta_3) = K(\tau, \eta_1)$. Since $K(\tau, \eta_3)$ is not dependent on η_1, the expression on the left in (31) must be a function of τ alone, $= K(\tau)$. From this is derivable

$$F(\tau + \eta_3) - K(\tau)F(\eta_3) = F(\tau + \eta_1) - K(\tau).$$

The expression on the left is independent of η_1, and the expression on the right is independent of η_3; therefore, both are dependent on τ alone:

$$F(\tau + \eta_3) - K(\tau)F(\eta_3) = G(\tau).$$

If in the foregoing equation η is substituted for η_3, the following functional equation for F is obtained:

$$F(\tau + \eta) = K(\tau)F(\eta) + G(\tau). \tag{32}$$

With $\tau = 0$

$$F(\eta) = K(0)F(\eta) + G(0);$$

therefore,

$$K(0) = 1 \qquad G(0) = 0.$$

On the other hand, with $\eta = 0$

$$F(\tau) = K(\tau)F(0) + G(\tau).$$

Since τ is arbitrary, we shall replace it once by $\tau + \eta$ and subsequently by η; we obtain

$$F(\tau + \eta) = K(\tau + \eta) \cdot F(0) + G(\tau + \eta),$$
$$F(\eta) = K(\eta) \cdot F(0) + G(\eta).$$

From (32) we obtain

$$K(\tau + \eta) \cdot F(0) + G(\tau + \eta)$$
$$= K(\tau)K(\eta)F(0) + K(\tau)G(\eta)G(\tau).$$

Since $F(0)$ is completely arbitrary, that is, without any possible influence upon the functions K and G, it follows that

$$K(\tau + \eta) = K(\tau)K(\eta),$$
$$G(\tau + \eta) = K(\tau)G(\eta) + G(\tau).$$

The first of these two functional equations has only two solutions according to Cauchy: first, the exponential function

$$K(\tau) = e^{C\tau}$$

and second, the constants 1 and 0. This trivial case becomes important, because it leads to the linear function for F as is seen easily by setting $K = 1$. For G the functional equation $G(\tau + \eta) = G(\tau) + G(\eta)$ is obtained which, according to Cauchy, results in the linearity of G and thus in the linearity of F. $K = 0$ leads to the case $F = 0$ which must obviously be excluded. The other important case is the exponential form for K. If we substitute it in the second of the functional equations, there results for G

$$G(\tau + \eta) = e^{C\tau}G(\eta) + G(\tau).$$

If τ and η are interchanged, we obtain

$$G(\tau + \eta) = e^{C\eta}G(\tau) + G(\eta).$$

From this

$$\frac{G(\tau)}{e^{C\tau} - 1} = \frac{G(\eta)}{e^{C\eta} - 1}$$

is immediately derivable. The left-hand side is independent of η, the right-hand side of τ. Therefore, both sides are equivalent to a constant A' and we obtain

$$F(\tau) = e^{C\tau}F(0) + A'(e^{C\tau} - 1).$$

If we set $F(0) + A' = A$, $- A' = B$, since $F(0)$ is an arbitrary constant, we obtain

$$F(\tau) = Ae^{C\tau} + B.$$

This constitutes the required proof. At the same time, K. Fladt has furnished a second, very elegant proof for the uniqueness of definition 12. It is seen easily that (17) is not satisfied for the exponential function. Therefore, the linear function remains as the only solution of conditions (15) to (17).

The system K', which is obtained by substituting e^{τ} for $F(\tau)$, has interesting properties. According to (6), (7), and (8), the transformations become

$$\left.\begin{aligned} \rho &= \tfrac{1}{2}[\log (t + r) - \log (t - r)] \\ \vartheta &= \varphi \\ \tau &= \tfrac{1}{2}[\log (t + r) - \log (t - r)]. \end{aligned}\right\} \tag{33}$$

A point resting in K' has constant ρ and ϑ; we obtain its equation of motion in K from the first equation if we set $\rho = constant$. In addition, we want to shift the initial time point in K, that is, replace t by $t + p$ where p is constant. Then we can derive

$$\left.\begin{aligned} r = r_0 q \cdot t + r_0 \qquad r_0 &= p \cdot \frac{e^{2\rho} - 1}{e^{2\rho} + 1} \\ q &= \frac{1}{p}. \end{aligned}\right\} \tag{34}$$

Viewed from K, the point is moving radially away from A with the velocity $r_0 \cdot q$. For $t = 0$, r_0 is its distance from A since r_0 is constant and q is the same for all points; therefore, the velocity of each point is constant and proportional to its initial distance r_0 from A. Seen from K, the system K' is in a state of uniform, radial expansion. Nevertheless, K' is at rest light-geometrically; axiom IV,1 is valid if clocks are established on the moving points according to τ time. The round-trip axiom IV,2 is also valid (cf. pp. 161–162). Furthermore, axiom III as well as theorems 16 and 17 are valid. Although no rigid bodies are used, it can be seen that K' is not uniquely rigid, because the light geometry becomes non-Euclidean.†

By means of axioms I to V, definitions 1 to 12, and theorems 1 to 18 the light geometry of a single system is complete. We have thus indicated a unique procedure of selecting a system of real points as a distinguished coordinate system and of constructing in it the temporal and spatial measure relations. Subsequently we shall investigate the transition to uniformly moving systems and the state of motion of this family.

† See editor's reference notes, p. 203.

§ 14. *Uniformly Moving Systems*†

Definition 13. A point is moving uniformly relative to a system K if it travels equal distances in equal time intervals, both measured in K.

Definition 14. A system K′ is moving uniformly relative to a system K if all points of K′ are moving uniformly with equal and equidirectional velocity relative to K.

K′ will consist of real points if *K* consists of real points and if every point of *K′* is moving relative to *K* with a velocity less than that of light. Then axioms I to III must be valid in *K′*, since they refer to all real points. In the following we shall consider only such systems *K′* which are moving uniformly relative to an inertial system. We wish to prove the following theorem.

THEOREM 19. In a system *K′* moving uniformly relative to an inertial system, axioms I to V are valid; it is therefore also an inertial system.

According to what has been said above, we need to prove the theorem only for axioms IV and V. We shall turn first to axiom IV, 1. We imagine clocks established in *K′* which are regulated so that they always show the same time as that clock of *K* which they are passing at the moment. We consider a signal $A'B'A'$ in *K′*; in *K* it travels the path $A_1B_1C_1$ (fig. 8). If we repeat the signal $A'B'A'$ at a later time, it will travel a

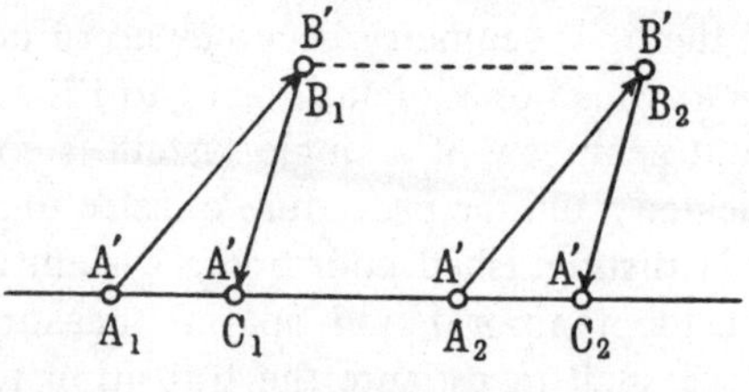

Fig. 8. Pertaining to the proof of axiom IV, 1 in a moving system.

† See editor's reference notes, p. 203.

path $A_2B_2C_2$ in K. Figure $A_2B_2C_2$, however, is merely shifted parallel to $A_1B_1C_1$, because the distances A_1A_2, B_1B_2, and C_1C_2 are equal and equidirectional according to the definition of uniform motion. Therefore, $\overline{A_1B_1C_1} = \overline{A_2B_2C_2}$ (theorem 18) measured in K. According to the chosen definition of time in K', $\overline{A'B'A'} = \textit{constant}$ for this system, too. Since this relation holds for any two points of K', axiom IV, 1 is valid in K'. But the chosen definition of time in K' is not necessary for this axiom; a linear function of this time would be sufficient.

After we have determined the time transformation from K to K' by $t = \tau$, we can also define the measurement of length in K' by reference to K. At a time point t every point of K' will coincide with a point of K, and thus a sequence of points l' in K' is coordinated to every segment l in K; l' is called a segment in K'. At another time point t_1 another segment l_1 in K will be coordinated to the segment l' defined above; but because of the uniform translation of all points of K', $l_1 = l$. Therefore, we can equate the lengths of l' and l. At a time point t the geometry of K' is identical with the geometry of K; therefore, it is Euclidean.[22] We call it *extraneous geometry* in K', because it is defined by a relation to an extraneous system. Between K and K' there exists the Galilean transformation (this is its strict definition):

$$\xi_i = x_i - v_i t \qquad \tau = t. \tag{35}$$

In K' the velocity of light $\mathfrak{p}$ is not constant, but determined by the vectorial addition

$$\mathfrak{p} = \mathfrak{c} - \mathfrak{v}. \tag{36}$$

By means of these expressions we can now prove theorem 19 for axiom IV, 2. First, we show that for the velocity of light obtained by the vectorial addition, axiom IV, 2 is valid. We consider a triangle ABD in K' (fig. 9; we are omitting the prime marks from the magnitudes of K' and are using them only to indicate the opposite sense of a given direction). We call the

[22] Axiom V is not proved thereby, because this axiom refers to a different definition of time and length. Cf. below.

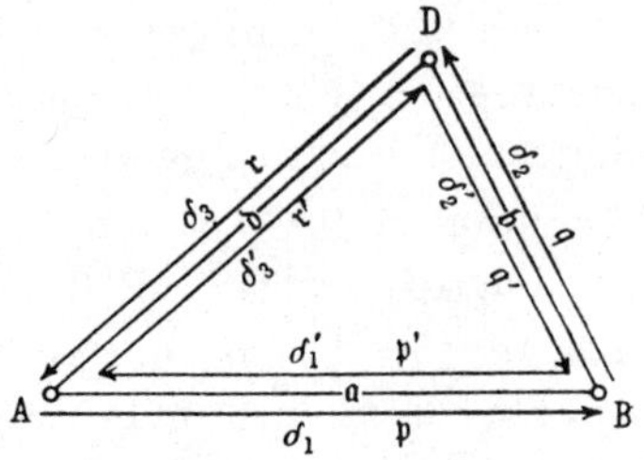

Fig. 9. Pertaining to the proof of the round-trip axiom IV, 2 in a moving system.

travel times $\overline{AB}$, $\overline{BD}$, $\overline{DA}$: $\delta_1\delta_2\delta_3$, and in the opposite direction $ADBA$: $\delta_3'\delta_2'\delta_1'$. We call the sides of the triangle $\mathfrak{a}$, $\mathfrak{b}$, $\mathfrak{d}$, where the direction of these vectors is assumed in the sense $ABDA$; let $\mathfrak{a}$ be the vector AB. The three velocities of light are called $\mathfrak{p}$, $\mathfrak{q}$, $\mathfrak{r}$, respectively, and for the opposite path $\mathfrak{p}'$, $\mathfrak{q}'$, $\mathfrak{r}'$. Their direction is determined by $\mathfrak{a}$, $\mathfrak{b}$, $\mathfrak{d}$. Thus, for the components of $\mathfrak{a}$, $\mathfrak{p}$, etc.:

$$\left.\begin{aligned} \mathfrak{a}_\nu &= \delta_1 \mathfrak{p}_\nu, & \mathfrak{b}_\nu &= \delta_2 \cdot \mathfrak{q}_\nu, & \mathfrak{d} &= \delta_3 \mathfrak{r}_\nu, \\ \mathfrak{a}_\nu &= -\delta_1' \mathfrak{p}_\nu', & \mathfrak{b}_\nu &= -\delta_2' \mathfrak{q}_\nu', & \mathfrak{d}_\nu &= -\delta_3' \mathfrak{r}_\nu'. \end{aligned}\right\} \qquad (37)$$

We determine $\mathfrak{p}$. We write (36) in components:

$$\mathfrak{p}_\nu = \mathfrak{c}_\nu - \mathfrak{v}_\nu = \frac{\mathfrak{a}_\nu}{\delta_1}.$$

We derive

$$\mathfrak{c}_\nu^2 = \left(\frac{\mathfrak{a}_\nu}{\delta_1} + \mathfrak{v}_\nu\right)^2$$

$$c^2 = \sum_\nu \mathfrak{c}_\nu^2 = \sum_\nu \left[\frac{\mathfrak{a}_\nu^2}{\delta_1^2} + \mathfrak{v}_\nu^2 + \frac{2}{\delta_1}\mathfrak{a}_\nu \mathfrak{v}_\nu\right] = \frac{a^2}{\delta_1^2} + v^2 + \frac{2}{\delta_1}(\mathfrak{a}\mathfrak{v}),$$

where c, a, and v are the absolute values of these vectors. Solved for δ_1, we obtain

$$\delta_1 = \frac{(\mathfrak{a}\mathfrak{v})}{c^2 - v^2} + A, \qquad (38)$$

where

$$A = +\sqrt{\frac{(\mathfrak{a}\mathfrak{v})^2}{c^2 - v^2} + \frac{a^2}{c^2 - v^2}}. \qquad (39)$$

The root is positive, because $\delta_1 > 0$. If $\mathfrak{p}'_v$ and δ'_1 are determined analogously, we obtain

$$\delta'_1 = -\frac{(\mathfrak{a}\mathfrak{v})}{c^2 - v^2} + A. \tag{40}$$

Analogous expressions result for δ_2, δ'_2... Thus

$$\delta_1 - \delta'_1 = \frac{2(\mathfrak{a}\mathfrak{v})}{c^2 - v^2}, \qquad \delta_2 - \delta'_2 = \frac{2(\mathfrak{b}\mathfrak{v})}{c^2 - v^2},$$
$$\delta_3 - \delta'_3 = \frac{2(\mathfrak{d}\mathfrak{v})}{c^2 - v^2} \tag{41}$$

and with $\delta = \sum \delta_i$:

$$\left.\begin{aligned} \delta - \delta' &= \delta_1 - \delta'_1 + \delta_2 - \delta'_2 + \delta_3 - \delta'_3 \\ &= \frac{2}{c^2 - v^2}\big((\mathfrak{a}\mathfrak{v}) + (\mathfrak{b}\mathfrak{v}) + (\mathfrak{d}\mathfrak{v})\big). \end{aligned}\right\} \tag{42}$$

The expression in the parentheses becomes

$$(\mathfrak{v}(\mathfrak{a} + \mathfrak{b} + \mathfrak{d})) = 0,$$

because

$$\mathfrak{a} + \mathfrak{b} + \mathfrak{d} = 0, \tag{43}$$

since these three vectors form a closed curve. It follows from (42) that

$$\delta = \delta', \tag{44}$$

that is, the validity of axiom IV,2 is proved.

We have derived axiom IV,2 for a certain definition of time. But the content of this axiom is independent of the definitions of simultaneity and time interval and therefore axiom IV,2 must be valid for every definition of time that does not destroy coincidences. Therefore, axiom IV,2 is generally valid in K'.

We shall now turn to axiom V. Because of the validity of axioms I to IV we can carry through definition 8 of simultaneity and definitions 9, 10, and 11 of spatial geometry in K'. Thereby the light geometry of K' is completely determined. We want to show that it becomes Euclidean, that is, that axiom V is valid. For this purpose we shall consider the transformation from the extraneous geometry in K' to the light geometry in K' which constitutes a change of the metrical determination. A straight line in the sense of the light geometry in K' is a sequence of points which a light signal reaches consecutively. Since the light as well as the points of K' are moving uniformly relative to K, at a time t these points, although not coinciding with the path of this light ray in K, lie nevertheless on a straight line. Straight lines in the extraneous geometry are therefore also straight lines in the light geometry, and for this reason the transformation from the extraneous geometry to the light geometry is linear.[23] This transformation consists in a multiplication of lengths in one linear direction by a constant factor κ.

Let $A'B'$ be an interval in the direction of motion, let $A'C'$ be perpendicular to it (this is defined in the extraneous geometry), let its coefficients of measure in the extraneous geometry be ξ_1 and ξ_2, and let $\xi_1 = \xi_2$. Then, according to the Galilean addition of velocities,

$$\overline{A'B'A'} = \frac{\xi_1}{c - v} + \frac{\xi_1}{c + v}, \qquad \overline{A'C'A'} = \frac{2\xi_2}{\sqrt{c^2 - v^2}}.$$

From this expression is derivable

$$\overline{A'B'A'} \cdot \sqrt{1 - \frac{v^2}{c^2}} = \overline{A'C'A'}. \tag{45}$$

(This is the familiar calculation of the Michelson experiment.) If the light-geometrical lengths of the two intervals are x_1' and x_2', then, according to definition 10, $x_1' \cdot \sqrt{1 - \frac{v^2}{c^2}} = x_2'$.

Light-geometrically, therefore, $A'B' > A'C'$. We are still free to define the unit of the light geometry. If we set

$$x_2' = \xi_2,$$

---

[23] It should be noted that this result is not yet derivable from the fact that light rays shall be straight lines in K and K'; this condition is also satisfied by the uniform radial expansion according to § 13. The condition must be added that the system K' is moving uniformly relative to K by definition.

then

$$x_1' = \kappa \cdot \xi_1, \qquad \kappa = \frac{1}{\sqrt{1 - \frac{v^2}{c^2}}}. \tag{46}$$

But if we set $x_1' = \xi_1$, then $x_2' = \frac{1}{\kappa} \cdot \xi_2$. Both expressions change the metrical determination in K' without destroying its Euclidean character; therefore, axiom V is valid for the light geometry of the uniformly moving system K'.

The foregoing is a proof of theorem 19 as well as a proof of the assertion that the point system K' specified by definition 14 is rigid according to definition 12. This assertion must be proved because definition 14 does not prescribe the motion of K' as a totality, but merely determines the motion of each particular point of K'. Conversely, it can be shown, too, that K is moving uniformly relative to K' according to definition 14. This assertion presupposes a metrical determination for K'; but it is valid both if the extraneous geometry is used in K'—then it is proved immediately when transformation (35) is solved for ξ_i, τ—and if the light geometry according to definitions 1 to 12 is carried through in K' itself. The assertion is proved when the transformations (47) and (52) of the following section are solved for $x_i' t'$. A proof of the general linear transformation is given in § 17, requirement 6.

§ 15. *The Lorentz Transformation*

The Lorentz transformation is that transformation which connects K and K' if in both systems the metrical determination is defined in light-geometrical terms. We shall have to derive it from our axioms and definitions. However, three more definitions must be added referring to the comparison of the units of K' with those of K.

Definition 15. Length ($l_{K'}^{K}$) *of a segment at rest in the moving system* K', *measured in* K, *is the distance in* K *between simultaneous positions of its end points, measured in* K.

This definition of the length of the moving segment must be added, because in definition 11 merely the length of the segment at rest is defined; only both definitions together make kinematics possible. Definition 15 is also necessary if the length of the segment at rest is not defined light-geometrically, but by means of rigid measuring rods. Occasionally the length defined by definition 15 has been called "an apparent length" in contrast to the "true length." This is a misuse of language. In both instances we are dealing with definitions for which the predicate "true" is meaningless. On the other hand, neither of them is dispensable for the description of spatiotemporal processes. What is the "true" diameter of the earth? The diameter from pole to pole or the diameter through the equatorial plane? In this example the meaninglessness of the predicate "true" is easily noticed. Both distances together determine the magnitude to be measured. At most one might ask whether one distance is longer than the other. This result is not dependent on these definitions alone, but also on the other definitions and axioms, that is, on empirical facts. For relativistic kinematics $l^K_{K'} < l^K_K$; this result follows from the definition of simultaneity and the axioms of the motion of light. It cannot be proved that $l^K_{K'} = l^K_K$; this equation can merely be made a requirement by means of which, conversely, a special simultaneity is defined (cf. § 27). Yet this requirement has nothing to do with uniqueness or truth.

Definition 16. The length of the unit at rest in K′, measured in K, has to be equal to the length of the unit at rest in K, measured in K′. $l^K_{K'} = l^{K'}_K$.

Definition 17. The unit of time in K′ is to be chosen in such a way that the velocity of light, related to the unit of length projected according to definition 16, receives the same numerical value in K′ as in K.

We shall now prove two auxiliary theorems concerning the transformations.

THEOREM 20. Two point events occurring in *K* perpendicular to the direction of motion of *K′* are simultaneous in *K′*.

Proof. Let A, B be two points in K perpendicular to the direction of motion (fig. 10). Let the points A', B' of K' coincide with them at the time t_0 (defined in K). At this moment each of them emits a light signal; these signals intersect at a point C' of the connecting line $A'B'$ at the time t_1. At this time, C' coincides with C of K. Since the velocity of light is constant in K, $AC = BC$. Since C' is moving on a line parallel to the direction of motion in K, this parallel (seen from K) must lie in the

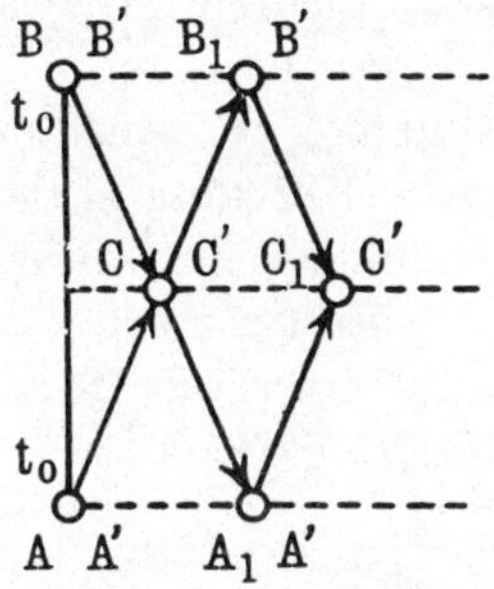

Fig. 10. Pertaining to theorem 20.

middle between the lines of motion of the points A' and B'. We must now show that C' lies in the middle of $A'B'$, measured in K', because then the departures were also simultaneous for K'. For this purpose we send two signals $C'A'C'$ and $C'B'C'$ from C' at the time t_1. When they are at A' or B', these points coincide with A_1 or B_1, respectively, in K. Since AA_1 is parallel to BB_1 and the line of motion of C is a parallel at equal distance from both, these signals intersect on this line, that is, in C'. Therefore, $\overline{C'A'C'} = \overline{C'B'C'}$ and according to definition 10 the point C' lies in the middle of $A'B'$.

THEOREM 21. A segment lying perpendicular to the direction of motion in K furnishes the same measurement of length in K'.

Proof. Let l_K be a segment at rest in K perpendicular to the direction of motion. Let its magnitude in K be l_K^K, in K' be $l_K^{K'}$. Let $l_{K'}$ be an arbitrary segment at rest in K'; let its magnitude in K' be $l_{K'}^{K'}$, in K be $l_{K'}^{K}$. Let us assume that $l_K^K > l_K^{K'}$;

then, according to definition 16, also $l_{K'}^{K'} > l_{K'}^{K}$. If we assume $l_{K'}^{K'} = l_{K}^{K'}$, that is, if we choose the (still available) segment $l_{K'}$ in K' identical with the measure of l_K in K', then, because of theorem 20, l_K is conversely the measure of this segment $l_{K'}$, measured in K, that is, $l_{K'}^{K} = l_{K}^{K}$. If the last two equations are substituted in the first inequality, then $l_{K'}^{K'} < l_{K'}^{K}$ results. This expression contradicts the second inequality. A corresponding contradiction arises if $l_{K}^{K} < l_{K}^{K'}$ is assumed. Therefore, $l_{K}^{K} = l_{K}^{K'}$ and $l_{K'}^{K'} = l_{K'}^{K}$, that is, theorem 21 is valid:

We can now easily derive the Lorentz transformation if we use the intermediate system ξ_i, τ, which describes K' in terms of the *extraneous geometry*. Because of theorem 21, only (46) remains among the possibilities of choosing a unit mentioned in § 14. If the lengths and times are called x' and t', the following expression is obtained by means of (35) and (46):

$$x_1' = \frac{x_1 - vt}{\sqrt{1 - \dfrac{v^2}{c^2}}} \qquad x_2' = x_2 \qquad x_3' = x_3 . \tag{47}$$

We shall now determine the time transformation. Because of theorem 20, t' depends only on x_1 and t. In K' a signal $A'B'A'$ is sent along the x_1' axis and back. Its times in τ time are $\tau_1\tau_2\tau_3$ and, therefore,

$$\tau_2 = \tau_1 + \frac{\xi_1}{c - v} = \tau_3 - \frac{\xi_1}{c + v} . \tag{48}$$

The adjustment of the t' time is accomplished in two steps. First, the simultaneity according to definition 8 is determined, that is, the periods of all clocks are made identical with those of the neighboring τ clocks; only the starting points of all clocks are changed. The resulting time is called t''. Thus

$$t_2'' = \frac{t_1'' + t_3''}{2} \qquad t_1'' = \tau_1 \qquad t_3'' = \tau_3 . \tag{49}$$

If t_1'' and t_3'' are eliminated and the index "2" is omitted, then from (48) and (49)

$$t'' = \tau - \frac{v\xi}{c^2 - v^2} = \frac{t - \frac{v}{c^2} \cdot x}{1 - \frac{v^2}{c^2}} \tag{50}$$

is derivable. Only now do we adjust the period in K' according to definition 17. Then a segment of the length 1 along the x_2' axis, run through in τ time (and therefore also in t'' time) in the time $\frac{1}{\sqrt{c^2 - v^2}}$ measured by means of t', must be run through in the time $1/c$; therefore, $\tau \cdot \lambda = t'$, and λ is given by $\frac{1}{\sqrt{c^2 - v^2}} \cdot \lambda = \frac{1}{c}$. Then

$$t' = \tau\sqrt{1 - \frac{v^2}{c^2}} = t''\sqrt{1 - \frac{v^2}{c^2}}. \tag{51}$$

With (50)

$$t' = \frac{t - \frac{v}{c^2} x}{\sqrt{1 - \frac{v^2}{c^2}}} \tag{52}$$

is obtained. Equations (47) and (52) are *Lorentz transformations.*

By means of the these transformations it is easily confirmed that definition 16 holds also for the segment lying in the direction of motion, that is, that definitions 16 and 10 do not contradict each other.

Other derivations of the Lorentz transformation are not essentially different from the one given above. In Einstein's[24] second derivation, definition 16 and theorem 21 are used as presuppositions; the additionally used linearity of the transformation was derived by us from the definition of uniform

---

[24] A. Einstein, *Relativity, the Special and the General theory: A Popular Exposition* (London: Methuen, 1920), Appendix.

motion and of the straight line. In Einstein's first derivation,[25] definition 16 is expressed by the requirement that for the contraction factor of the lengths $\varphi(v) = \varphi(-v)$ (in the notation of Einstein's publication). Theorem 21 is derived by means of definition 10 contained in the principle of the constancy of the velocity of light.

It is remarkable that the Lorentz transformation, conceived in light-geometrical terms, does not contain a new axiom but depends solely on definitions and axioms of the single system. With respect to light, there does not exist a special axiom of uniform translation. Our derivation of the Lorentz transformation via the Galilean transformation has shown that it is merely a *reorganization* of the relations of uniformly moving systems contained in the Galilean transformation *into a new metrical determination.* Physically new is Einstein's idea that material structures do not adjust to the Galilean metrical determination but to the light-geometrical one. This result cannot be derived from definitions but requires new axioms. We shall formulate them in the matter axioms.

§ 16. *The State of Motion of Inertial Systems*†

Until now we have proved two properties of the inertial system defined in light-geometrical terms: first, that it is uniquely rigid, and second, that a system moving uniformly relative to an inertial system is also an inertial system. We must now prove a third property: that the state of motion of the inertial system is determined up to a uniform translation, that is, that the class obtained by means of the second property contains all systems in which axioms I to V are valid.

For this purpose we need to prove only that linear transformations must exist between two systems in which the light geometry has been carried through. The use of the definitions of the order of time (definition 1), of the measure of time (definition 7), of simultaneity (definition 8), of the straight line

[25] *Ann. d. Phys.* vol. 17 (1905), p. 891.

† See editor's reference notes, p. 203.

(definition 9), and of the equality of segments (definition 10), in addition to the axioms for the characterization of the inertial system, does not constitute a restriction of generality. These definitions can be carried through in every system in which axioms I to V are valid; they do not constitute a change in the selection of the corresponding real points. If the transformation from K to K' is necessarily linear according to these definitions, then the states of motion of the systems relative to one another are determined. Nevertheless, it is due to these definitions carried through in K that we call the state of motion of K' relative to K uniform. If another metrical determination were used in K, the motion of K' relative to K would be expressed in more complicated terms. But this consideration concerns only the designation of the state of motion which is not possible without a definition of the metrical determination. That selection from all real points moving in any way, which can be combined in the system K', is not affected thereby.

A system K in which axioms I to V are valid, and in which the above-mentioned definitions are carried through, is characterized by the metrical determination

$$ds^2 = dx_1^2 + dx_2^2 + dx_3^2 - dx_4^2 = [dx_\nu^2]. \tag{53}$$

The bracketed expression is an abbreviation for the special summation in which the fourth term receives a negation sign. The motion of light is given by $ds^2 = 0$. If we now go to a system K' in such a way that in K' the motion of light is also given by $[dx_\nu'^2] = 0$, then axioms I to V are equally valid in K'. The most general transformations that can connect systems with Euclidean light geometry are those which transform $[dx_\nu^2] = 0$ into $[dx_\nu'^2] = 0$. This result requires that $[dx_\nu^2]$ be identically transformed into $\lambda \cdot [dx_\nu'^2]$ by the transformation, where $\lambda = \lambda(x_1' \ldots x_4')$. The system K' obtained in this manner does not yet have the metrical determination specified by the light geometry; but if definition 7 of the measure of time and definition 10 of the equality of distances are introduced, that is, if ds^2 is divided by λ, then a metrical determination ds'^2 of the form (53) results.

From a mathematical point of view, the problem of these transformations has long been settled. In addition to the linear transformations, the most general among such transformations are the spherical transformations which transform spheres into spheres, for example,

$$x_i = \frac{x'_i}{r'^2} \qquad r'^2 = [x'^2_i]. \tag{54a}$$

But all of these linear transformations have a singularity; they transform a finite point into infinity and vice versa. This means that in such systems as K' there exists a point for which axioms I to V do not hold. If therefore the requirement is made that the light axioms are to be valid for all points of a system, only the linear transformation remains, that is, the state of motion of such a system is determined up to a uniform translation.

Let us consider the above-mentioned transformation with reciprocal radii; its inverse transformation is

$$x'_i = \frac{x_i}{r^2} \qquad r^2 = [x_i^2] \qquad r^2 = \frac{1}{r'^2}. \tag{54b}$$

A point B' with the coordinates $x'_1 = b$, $x'_2 = x'_3 = 0$, at rest in K', has in K the equation of motion $b = \dfrac{x_1}{x_1^2 - x_4^2}$; therefore,

$$x_1 = \frac{1}{2b} + \sqrt{\frac{1}{4b^2} + x_4^2} \qquad \dot{x}_1 = \frac{x_4}{\sqrt{\dfrac{1}{4b^2} + x_4^2}}, \tag{55}$$

where $\dot{x}_1 = \dfrac{dx_1}{dx_4}$ symbolizes the velocity in K. The point thus moves radially away from the starting point A in K while its velocity increases and approaches asymptotically the velocity of light 1. At the time $x_4 = 0$ the point is at the distance $1/b$ from A. Previously it moved in the opposite direction and traveled from the outside toward A. At the moment $x_4 = 0$ it reverses its direction of motion. All points of K' perform such radial motions relative to A; the reversal of the motion starts at the time $x_4 = 0$. No two points of K', measured in K, have a fixed distance from each other (nor can they be connected by rigid rods—albeit this fact does not belong to the light geometry). Nevertheless, all light axioms are valid for these moving points. The distance between two such points may be called constant for K' and the metric of time may be adjusted correspondingly. This result constitutes the change of the metrical determination in K', that is, the division of ds^2 by λ. This result is similar to the uniform expansion discussed in § 13; but in this case of an accelerated expansion the spatial geometry becomes Euclidean.

An essential difference from the uniform expansion considered earlier consists in the fact that singularities occur. If $b = 0$, then for every x_4, $x_1 = \infty$, that is, the starting point A of K', seen from K, lies in infinity. On the other hand, seen from K', A lies in infinity. It follows that not all points of K' can be reached by light signals.

Let B' and $\bar{B}'$ be two points at rest in K' on the x_1' axis with the coordinates $x_1' = b$ and $\bar{x}_1' = \bar{b}$. Let a light signal be sent from B' at the time x_4' which arrives at $\bar{B}'$ at the time $\bar{x}_4'$. We shall now describe this process in K. The two points are moving on the x_1 axis toward the outside; let $\bar{B}'$ be the exterior of the two. The departure of the light signal from B' occurs at the point x_1x_4, the arrival at $\bar{B}'$ at the point $\bar{x}_1\bar{x}_4$. Then

$$x_1 + (\bar{x}_4 - x_4) = \bar{x}_1,$$

since the distance $\bar{x}_1 - x_1$ is traveled by the light signal in the time $\bar{x}_4 - x_4$. If here the value of $\bar{x}_1$ is introduced from (55), then $\bar{x}_4$ can be calculated as a function of x_1, $\bar{b}$, x_4. Thus

$$\bar{x}_4 = \frac{-(x_1 - x_4)^2 + \dfrac{1}{\bar{b}}(x_1 - x_4)}{2(x_1 - x_4) - \dfrac{1}{\bar{b}}}. \tag{56}$$

For $\bar{b} = \dfrac{1}{2(x_1 - x_4)}$ the denominator vanishes and $x_4 = \infty$. There exists therefore a certain point $\bar{B}'$ that cannot be reached from B' by means of a light signal. Although the light signal travels faster than $\bar{B}'$ and the distance of the signal front from $\bar{B}'$ (measured in K) decreases continuously, $\bar{B}'$ will be overtaken only after an infinite time. This is a case for which axiom II, 1 does not hold.[26] B' as well as $\bar{B}'$ are real points moving with a velocity below that of light. But since the velocity of $\bar{B}'$ approaches the velocity of light asymptotically, $\bar{B}'$ is never overtaken within a finite time. At the time x_4, $\bar{B}'$ in K has the position $\bar{x}_1^*$. Therefore, since

$$\bar{x}_1^* = \frac{1}{2\bar{b}} + \sqrt{\frac{1}{4\bar{b}^2} + x_4^2} \quad \text{and} \quad \bar{b} = \frac{1}{2(x_1 - x_4)},$$

it follows that

$$\bar{x}_1^* = (x_1 - x_4) + \sqrt{(x_1 - x_4)^2 + x_4^2}.$$

If $x_4 = 0$, then $\bar{x}_1^* = 2x_1$. If a signal is sent from B' at the time $x_4 = 0$, it will overtake a point located at this time at twice the distance from the starting point A only after an infinite time. Still less will points lying farther outside ever be overtaken.

The process looks different in K'. Here the two points have the constant coordinates b and $\bar{b}$. Since the velocity of light in K' also $=1$, $\bar{x}_4' - x_4' = b - \bar{b}$ has a finite value. This result holds even if the time $\bar{x}_4 - x_4 = \infty$, seen from K. Transformation (54b) furnishes a finite $\bar{x}_4'$ for the infinite $\bar{x}_4$ and correspondingly a finite $\bar{x}_1'$ for the infinite $\bar{x}_1$.

From (54b) together with (55) it is possible to derive

$$\bar{x}_4' = \frac{\bar{x}_4}{\bar{x}_1^2 - \bar{x}_4^2} = \frac{\bar{x}_4}{\dfrac{1}{2\bar{b}^2} + \dfrac{1}{\bar{b}}\sqrt{\dfrac{1}{4\bar{b}^2} + \bar{x}_4^2}}.$$

If $\bar{x}_4 = \infty$, then $\bar{x}_4' = \bar{b}$, that is, it is finite. Furthermore, from

$$x_1 = \frac{1}{2b} + \sqrt{\frac{1}{4b^2} + x_4^2}$$

[26] However, axiom II, 2 is valid. It is possible to send an earlier signal from B' that will reach $\bar{B}'$ at a given time. This fact holds also for points lying outside of $\bar{B}'$; they can be reached by earlier signals from B'.

and the vanishing of the denominator in (56), there results

$$x_4 = \frac{2\bar{b} - b}{4\bar{b}(b - \bar{b})}.$$

Finally, from these two equations we obtain

$$x_4' = \frac{x_4}{x_1^2 - x_4^2} = 2\bar{b} - b$$

and, therefore, $\bar{x}_4' - x_4' = b - \bar{b}$.

As a consequence of this mapping of an infinite time interval of K upon a finite one in K', contradictions arise for an observer in K'. He calculates that the signal from B' will reach $\bar{B}'$ after a finite time; he discovers, however, by his experiments that it never reaches $\bar{B}'$. Still less will points lying farther outside ever be reached by the signal. The observer discovers that the light geometry can be carried through only in a finite domain around $\bar{B}'$; farther outside axiom II, 1 fails.

The singularity of transformation (54*a*) is also the reason we did not obtain this system in § 13.† System K' cannot be obtained by means of a time transformation at A, because all signals $AB'A$ are misrepresented in K'. K' can be constructed only from another starting point C', but this starting point will have a state of motion different from that of any point of K. The proof given in § 13 shows that from a fixed starting point only a single system can be constructed in which axioms I to V are valid. Yet from this fact nothing can be concluded concerning the state of motion of the starting point. If the starting point is given a certain nonuniform motion relative to the constructed system K, then in correspondence with our example another light-geometrical Euclidean system will exist around it.

An observer carrying through the light geometry in K' will believe at first that he is in an inertial system. Only after he has sent signals beyond a certain domain will he notice

† See editor's reference notes, p. 203.

the singularity; he must now conclude that he is not in an inertial system. It is characteristic solely of the inertial systems that the light axioms hold without singularities. The following objection might be made. If the observer is situated in an inertial domain, he will always be able to measure only a finite domain; he will therefore never know for certain whether or not singularities occur farther outside. This objection can be answered in the following way. The observer can search systematically for singularities. He assumes, at the outset, that his system K is an inertial system and then constructs a system K' according to (54*a*), calculates the limit at which singularities would have to occur, and tests this result by means of experiments. If he finds the singularities at the respective places, then his system K was an inertial system. If he does not find the singularities, then K' is the inertial system and K the expanded system. From this result he can calculate the place at which the singularities occur in K.

If the validity of axioms I to V is required for the whole system, without restriction to certain finite domains, then the light geometry is sufficient for the definition of the inertial system. If such a criterion based on the existence of singularities is to be avoided, material structures can be used. The validity of the light axioms I to V and of the matter axiom VII will uniquely determine the inertial system, even though this validity is tested only in finite domains.[27]

[27] H. Weyl, in the Appendix to his *Space-Time-Matter* (London: Methuen, 1922), uses a third method. The systems K and K' of our example can be distinguished by the fact that only in K a mass point free from the action of force travels in a straight line; in K' it would describe a complicated path. For this definition of the inertial system the concept *free from the action of force* is necessary. Therefore, a definition must be added that corresponds to our definition of "closed system" given in § 18; "free from the action of force" means "free from the action of physical forces" (definition 20). "Rectilinearity" is no longer the definition of "freedom from the action of forces," because "rectilinearity" is, conversely, defined in terms of "being free from the action of forces." The definition of the inertial system by means of mass points free from the action of forces contains therefore epistemologically the same problems as the definition by means of rigid bodies or natural clocks.

§ 17. *Summary of the Requirements Contained in the Lorentz Transformation*†

The Lorentz transformation results from the general point transformation by means of the following additional requirements.

Requirement 1. The transformation is required to transform a system in which the Euclidean light geometry is carried through into a system moving uniformly relative to the first in which the light geometry is also carried through. This requirement leads to the linearity of the transformation according to § 14.

The above requirement may be replaced by another one: the transformation is to connect those systems in which the light axioms hold without singularities. This requirement, too, leads to linearity (§ 16).

The most general linear transformations are

$$x_i = a^i_k x'_k + b^i. \tag{57}$$

They contain twenty constants. Three more constants are added: first, the constants c and c' which achieve the transition from x_4 and x'_4 to t and t', respectively, that is, to the specific time unit in K and K', respectively; second, a constant k occurring in the first ten restricting equations for the a^i_k and referring to the comparison of the units in K and K', as will be shown presently. The requirements given here dispose of these twenty-three constants.

From requirement 1 it follows, furthermore, that the transformation is required to transform $[dx^2_\nu]$ into $k^2[dx'^2_\nu]$ where k is an arbitrary constant. The following equations can be derived:[28]

$$a^m_i a^m_l = \delta_{il} k^2, \quad \text{but} = -\delta_{il} k^2 \quad \text{if } i = l = 4. \tag{58}$$

† See editor's reference notes, p. 203.

[28] According to the customary notation, we are omitting the summation sign; instead we are using the rule that it is necessary to sum over every index that appears twice. The symbol "δ_{il}" is defined by: $\delta_{il} = 1$ for $i = l$, $\delta_{il} = 0$ for $i \neq l$.

We call them *pseudo-orthogonality restrictions* or, if no confusion arises, briefly *orthogonality restrictions* when requirement 2 is added (ten equations).

Requirement 2. $k = 1$. This is a definition concerning the comparison of the units of K with those of K'; it corresponds to definition 16. If k receives a different value, the Euclidean light geometry is still valid in every system; merely the relativity restriction of definition 16 between the units is missing. See § 23 for such a transformation (one equation).

Requirement 3. c is to be determined by the choice of the units of time and length in K, the unit of length perhaps in terms of the standard meter and the unit of time in terms of the revolution of the earth. Frequently c is set $= 1$; this is the definition of a connection between the unit of length and the unit of time (one equation).

Requirement 4. $c = c'$. Between the units of length and time in K the same relation with respect to light is to exist as exists between the units of length and time in K'. This requirement corresponds to definition 17. See § 23 for a different requirement for c' (one equation).

Requirement 5. $b^i = 0$. The zero points of K and K' are to coincide (four equations).

Requirement 6. The three components of the velocity v of K' relative to K are to assume given values v_α ($\alpha = 1, 2, 3$). This requirement furnishes the following three equations for the a^i_k:

$$\frac{a^\alpha_4}{a^4_4} = v_\alpha \qquad (\alpha = 1, 2, 3)\,. \tag{59}$$

It is customary to set $v_2 = v_3 = 0$, $v_1 = v$.

Calculation. We determine the equation of motion for a point P' at rest in K' with the spatial coordinates $x'_\alpha = p'_\alpha$. From (57) it follows that

$$x_\alpha = a^\alpha_\beta p'_\beta = a^\alpha_4 x'_4 + b^\alpha \qquad \alpha, \beta = 1, 2, 3\,.$$

The fourth equation yields

$$x_4 = a^4_\beta p'_\beta + a^4_4 x'_4 + b^4$$

$$x'_4 = \frac{1}{a^4_4} x_4 - \frac{a^4_\beta}{a^4_4} p'_\beta - \frac{b^4}{a^4_4}.$$

If these expressions are substituted in the preceding equation, we obtain x_α explicitly represented by x_4 and $p'_1 p'_2 p'_3$, namely, the desired equation of motion:

$$x_\alpha = \frac{a^\alpha_4}{a^4_4} x_4 + \frac{A^{\alpha 4}_{\beta 4}}{a^4_4} p'_\beta + b^\alpha - \frac{a^\alpha_4}{a^4_4} b^4 \qquad \alpha, \beta = 1, 2, 3, \quad (60)$$

where $A^{\alpha 4}_{\beta 4}$ designates that subdeterminant of second order of the a^i_k which is constructed by means of the horizontal rows α, 4 and the vertical rows β, 4. The coefficient of x_4 is the velocity of P' measured in K. The coefficient is independent of the p'_β; the linear transformation satisfies therefore definition 14 of the uniform motion of a system.

Requirement 7. The three spatial axes X'_α of K' are to be arranged so that their position in K, determined for a time point in K, is parallel to X_α in K.

This requirement furnishes for the a^i_k the equations

$$A^{\alpha 4}_{\beta 4} = 0 \qquad \begin{array}{c} \alpha, \beta = 1, 2, 3 \\ \alpha \neq \beta. \end{array} \qquad (61)$$

Of the resulting six equations three are consequences of the others because of the pseudo-orthogonality restrictions. Hence this requirement furnishes only three equations.

Calculation. Let a point of the X'_2 axis have the coordinates $x'_2 = p'_2$, $x'_1 = x'_3 = 0$. Its spatial coordinates in K at the time x_4 are given according to (60) by

$$x_\alpha = \frac{a^\alpha_4}{a^4_4} x_4 + \frac{A^{\alpha 4}_{24}}{a^4_4} p'_2 + b_\alpha - \frac{a^\alpha_4}{a^4_4} b_4.$$

Requirement 7 demands that all points of the x'_2 axis at the moment x_4 have the same coordinates x_1 and x_3; x_1 and x_3 must therefore be independent of p'_2. This requirement furnishes

$$A^{14}_{24} = 0, \qquad A^{34}_{24} = 0.$$

Analogous results are obtained for the other axes.

The best method to show that these equations are not independent is to use the imaginary x_4 and to assume the orthogonality of the a^i_k. Always two of the thirty-six subdeterminants A^{il}_{km}, arranged in pairs, are equal, namely, those which serve as complements of the determinant of the a^i_k. (Cf. M. von Laue, *Das Relativitätsprinzip*, Braunschweig: Vieweg, Vol. I [1913], 64–65.) Orthogonality restrictions exist also for the remaining eighteen magnitudes A^{il}_{km}, namely, six equations for the squared sums and fifteen equations for the mixed sums, altogether twenty-one equations. Six of these equations are consequences of the others because of the equality of the pairs of the A^{il}_{km}, namely, three of the squared sums and three of the mixed sums. The orthogonality restrictions thus furnish fifteen independent equations for the eighteen magnitudes A^{il}_{km} so that three of them are still available. This is the reason that three of the equations (61) determine the three others as consequences of the orthogonality restrictions, that is, of requirement 1. If three of the magnitudes (61) are set equal to zero, the other three magnitudes become zero, too. This result can be seen from the fact that the vanishing of the six magnitudes (61) does not contradict the orthogonality restrictions.

The seven requirements mentioned above furnish twenty-three equations by means of which the twenty-three constants of the transformation are determined.

§ 18. *Material Metrical Objects*

We shall now consider a second group of axioms referring to the metrical behavior of material objects. Only *rigid rods* and *natural clocks* represent such material objects. We must first define what is meant by them.

Definition 18. A natural clock is a closed periodic system; its unit is one period.

Definition 19. A rigid rod is a solid rod that is closed against all exterior forces.

We assume the meaning of "periodic system" and "solid rod" as defined by the language of physics. The only problem arises from the twice-used addition "closed." What does "closed" mean?

It is the problem of the term "changed." Is a rod called "changed under the influence of unknown forces" if the geometry measured by it becomes non-Euclidean? Not always; for instance, not on the surface of the earth which furnishes a two-dimensional non-Euclidean geometry. Nonetheless, such a definition might be given and the surface of the earth be called plane.[29] The deviations from plane Euclidean geometry would then have to be ascribed to a force *X* that changes the measuring rod and would show the following properties in experiments.

1. The force *X* affects all materials in the same way.
2. There are no insulating walls.

An insulating wall is a covering of any kind of material which does not act upon the enclosed rod with forces contradicting criterion 1. The assertion that the earth is a sphere is equivalent to the assumption that forces of the kind *X* do not exist.

An analogous problem exists for the measurement of a three-dimensional space. If the requirement is retained that the space be Euclidean, it might be necessary under certain circumstances to introduce a force of type *X*. Instead, the space might be called non-Euclidean, whereupon the force *X* will disappear. The latter conception is simpler in the sense of *descriptive* simplicity (§ 2). Therefore, the decision between the two descriptions is arbitrary. But it should be realized that a decision in favor of the first conception has the consequence that no definite geometrical form can be asserted; for instance, it would be equally justified to maintain that the earth is plane or oval.

---

[29] Cf., for instance, the presentation by R. Carnap, *Der Raum*, in *Kantstudien*, Erg.-Heft 56, p. 47. I have discussed the problem of the force *X* in detail in "La signification philosophique de la théorie de la relativité," *Revue Philos. de la France* (juillet 1922), pp. 32–40.

We call forces of type *X metrical forces*† because they depend on the choice of the metric. All other forces are called *physical forces* (e.g., heat, elastic and electric forces). Gravitation is one of the metrical forces. These considerations indicate the significance of Einstein's theory of gravitation. We shall now give two definitions.[30]

Definition 20. A system is called closed if no exterior physical forces act upon it.

Definition 21. The length of a rigid rod or of the period of a natural clock is that length which results when all physical forces are eliminated (either practically or by calculation); metrical forces are disregarded.

§ 19. *The Matter Axioms*

The matter axioms can be divided into two groups. Axioms VI refer to relations between material objects alone; the remaining axioms refer to relations between these objects and light.

Axiom VI, 1. Two rigid rods, which are at rest and of equal length at a given place, are of equal length, when measured by the same method, at any place and at any time when transported thereto as closed systems along different paths.

Axiom VI, 2. Two natural clocks, which are at rest at the same place and whose units are equal in a neighborhood comparison, have equal units when measured by the same method at any place and at any time when transported thereto as closed systems along different paths.

The axioms include the instance in which two rods or clocks are continuously at rest side by side. Their relative rest may be imagined to be judged by means of a rigid rod. The possibility of measuring with rigid rods and natural clocks depends on

---

† See editor's reference notes, p. 203.

[30] The definitions of this section refute the view that a definition of "rigid body" is impossible without reference to a metric, and answer at once a question put to me by Mr. Dingler (*Phys. Zeitschr.*, vol. 23 [1922], p. 52). These definitions use none but qualitative physical properties.

these axioms. This geometry is not complete, however; "straight line" and "simultaneity" must be defined by means of light.

The following axioms do not refer to arbitrary measurements but only to measurements in (light-geometrically defined) inertial systems. The validity of axioms VII to X is asserted for the time being for inertial systems only. (Later we shall extend the axioms to other systems; cf. § 33.) Yet it is asserted, at the same time, that always when axioms I to V are satisfied in a system, axioms VII to X will also be satisfied in this system. (The proof will be given in § 36.) The following axioms assert the identity of the geometry of rigid rods and natural clocks with the light geometry.

Axiom VII. Points at rest relative to one another according to definition 12, and only such points, can be connected by rigid rods.

Axiom VIII. Two intervals AB and AC, which are equal when measured by rigid rods, are also light-geometrically equal, when measured according to definition 10.

Axiom IX. The unit transported by rigid rods into a uniformly moving system is equal to the unit transferred according to definition 16.

Axiom X. The time unit of natural clocks is always of such a kind that the time $\overline{ABA}$ *of a light signal measured by these clocks at A is the same everywhere if AB is the same when measured by rigid rods.*[31]

---

[31] We have to add the following requirements to the formulations of axioms VIII and X. These axioms, going beyond axioms VI, are supposed to add a further property to the rods and clocks. For this purpose the independence of the axioms must be preserved. Axioms VI must not become consequences of VIII and X. Therefore, the antecedent of axiom VIII must be understood hypothetically: "If it is possible to measure two distances as equal independently of the choice of the rigid rods, then they will also be equal light-geometrically." The relation asserted in the axiom holds only if the antecedent holds. The antecedent (which presupposes axiom VI, 1) can therefore not be derived from the axiom. The formulation "which are equal when measured by rigid rods" is supposed to convey the hypothetical supposition. Corresponding considerations hold for axiom X. "The time unit of natural clocks" exists only if axiom VI, 2 is valid.

These axioms complete the philosophy of space and time of the special theory of relativity. The principle of the constancy of the velocity of light and the Lorentz transformation do not make statements about light alone but also about the *adjustment* of rigid rods and natural clocks; these material things adjust to the light geometry.

This result answers at once the problems concerning the *reality* of lengths of rods and times of clocks, the contraction of rods, and the retardation of clocks in Einstein's theory. The word "reality" is quite ambiguous in this context; but we can formulate clearly what is meant by the question of reality. The question means whether material objects adjust by themselves, that is, without human interference, to the requirements of the Lorentz transformation. The theory of relativity answers this question affirmatively. Only simultaneity has to be established deliberately once, that is, the starting point of clocks has to be set according to definition 8. But the retardation of clocks during the transport occurs spontaneously. In this sense the lengths of rods and times of clocks measured according to Einstein's conditions are real. This idea is formulated by the matter axioms.

The idea that the relations of the classical transformation, upon uniformly moving systems—the Galilean transformation (35)—are somehow distinguished complicates the understanding of this assertion. It is believed that material objects must necessarily adjust to these transformations as long as deliberate interventions or special forces do not produce a different effect. As a consequence some people believe that Einstein's kinematics concerns fictions; according to this conception, clocks and measuring rods must be imagined to be changed artificially in order to furnish the relations of Einstein's theory. Others who admit that clocks and rods adjust according to Einstein's theory search for special forces to which the deviation from the classical adjustment may be ascribed.

Both conceptions commit the error of regarding the classical transformation as a special one. The classical as well as the relativistic transformation can be defined light-geometrically (cf. §§ 14 and 15). It cannot be predicted a priori, however, which of them represents also the transformation relations of material objects. It is incorrect therefore if the first conception denies a priori the possibility of an adjustment to the relativistic transformation. The same a priori belief leads the second conception to an incorrect application of the requirement of causality. The second conception maintains that a special *cause* must be discovered for the *deviation* from the classical adjustment. This is a mistake. If the question concerning a cause is asked at all, then a cause must be sought for the adjustment of material objects to certain transformation relations to be defined light-geometrically. But this question concerns the adjustment to the classical as well as to the relativistic transformation. The word "adjustment"† used by Weyl for the first time in this context characterizes the problem very well. It cannot be an accident that two measuring rods are equal at every place in a neighborhood comparison; this fact must be explained as an *adjustment to the field* in which the measuring rods are embedded like test bodies. Just as the compass needle adjusts itself to the magnetic field of its immediate environment, even though merely in its direction, so measuring rods and clocks adjust their *unit lengths* to the *metrical field.* All metrical relations between material objects must be explained in this fashion. The answer can be given only by a detailed theory of matter of which nothing is known until now. It must be explained why the accumulations of certain parts of the field of special density, the electrons, provide a simple indication for the metric of the surrounding field. The word "adjustment," therefore, formulates a task rather than a solution. The existing state of affairs is formulated rigorously by the matter axioms without the use of the word "adjustment." Once we possess a theory of matter, we shall be able to explain the metrical behavior of material things. But it constitutes an unfounded a priori belief to regard the search for a cause of the deviation from classical metrical conditions as necessary.

† [Einstellung.]

Chapter Two

Critical Considerations

§ 20. *The Experimental Confirmation of the Axioms*

The axiomatic presentation is the best method for investigating the question which assertions of the theory of relativity have been confirmed experimentally. Once an axiomatic system has been constructed, only the experimental confirmation of the axioms needs to be tested. Since it is known what axioms are presupposed by a certain assertion, the empirical truth of such an assertion is determined at the same time. On the other hand, if it is discovered that a certain consequence is not confirmed, it is possible to indicate those axioms which have been violated and those axioms and theorems which remain unaffected. A necessary presupposition is the mutual independence of the axioms, an assumption which is an essential property of an axiomatic system.

Axioms I to II, 4 and axiom II, 6 express merely facts known before the theory of relativity. Only axiom II, 5 and axiom III are new. These axioms formulate the limiting character of the velocity of light. This assertion must not be regarded as a consequence of the theory of relativity; it is one of its presuppositions which can be tested experimentally. Since the investigations of the motion of electrons have shown that the kinetic energy of moving material particles becomes infinite at the velocity of light, the experimental evidence can be accepted as a fact. According to our definition of "direct signal," axiom III contains also the assertion of theorem 11 that the velocity of light is independent of the state of motion of the source. This result has also been confirmed experimentally and holds even in traditional optics.

Axiom IV, 1 contains the assertion that the velocity of light does not vary in time. This is also a familiar optical experience.

Furthermore, axiom IV, 2 is not an assertion peculiar to the theory of relativity. It holds in traditional optics because, as we showed in §14, it holds also on the assumption of the classical addition of velocities. New experiments that were to test this axiom would not be suitable for a decision between classical and relativistic optics. The same is true for axiom V which also holds in classical optics.

The light geometry is based on axioms I to V alone; it may therefore be regarded as empirically confirmed that this metrical determination can be carried through. Simultaneity, unit of length for the various directions, and unit of time distinguish the light geometry from the classical metrical determination; the differences are merely definitional. These concepts could also be defined light-geometrically in such a way that they furnish classical metrical relations. We have carried through these definitions (as an intermediate step) in §14.

We shall now discuss the matter axioms. Axioms VI, 1 and VI, 2 are customary presuppositions of measuring with rods and clocks that have been adopted by the theory of relativity. Axiom VII is also a traditional assumption expressing, in combination with the light axioms, the uniformity of the motion of light.

Axiom VIII is the immediate formulation of the facts observed in the Michelson experiment without any interpretation. It represents an inductive generalization, because the experiment could be carried through for one inertial system only (the system connected with the earth); yet the theory of relativity accepts the results of the experiment for all inertial systems. This decision is certainly in line with the usual inductive method of physics. Just because the axiom merely states the Michelson experiment, we can decide what consequences may or may not be drawn from this experiment. We shall come back to this matter in §§ 21 and 24.

Axiom IX cannot yet be confirmed directly. But indirect evidence speaks for this axiom. The axiom has nothing to do with the constancy of the velocity of light, but it would constitute a contradiction to the generally valid principle of relativity if this axiom were not satisfied. In addition, theorem 21, applied to rigid rods as coordinates, would not be valid; this appears to be improbable. Axiom IX could be replaced by an axiom corresponding to theorem 21; this would be logically equivalent to axiom IX. Theorem 21 may thus become relevant for an experimental test.

Axiom X has not yet been confirmed either. But the principle of relativity speaks for its validity. A demonstration of the transverse Doppler effect would be important for a confirmation of axioms IX or X (cf. § 24).

§21. *Remark on the Principle of the Constancy of the Velocity of Light*†

There exists a widespread misconception concerning the logical status of the principle that we shall call the *light principle.* Some people call it a statement that can never be confirmed but that nevertheless must be ranked at the head of the theory; others regard it as an empirical law like any other physical law; and still others believe that although the principle could never be confirmed positively, contradictions would arise if it were false. It is easy to deal with these questions on the basis of an axiomatic presentation. The light principle is not a single statement but a group of assertions, and more or less subject matter may be subsumed under this name. Moreover, it contains empirical statements as well as definitions and it becomes understandable why judgments concerning such a principle vary to such a degree.

If the light principle is conceived purely light-geometrically,

---

† See editor's reference notes, p. 204.

it contains the light axioms I to V and the definitions 1 to 17. On the basis of this conception, the empirical part of the principle, the axioms, may be regarded as well established. All that is new is merely the statement that light is the fastest signal. In order to derive the constancy of the velocity of light from these empirical facts, definitions must be added.

If the principle is interpreted as a statement not about light alone but about clocks and measuring rods, two assertions can be distinguished. The first, more restricted assertion says that the velocity of light is constant within an inertial system if the units are defined by rigid rods and natural clocks. This assertion contains axioms I to VIII and definitions 1 to 12 and 18 to 21. It is likewise well confirmed since the only axiom which did not hold in traditional optics, axiom VIII, has been confirmed by the Michelson experiment.

The wider assertion says that the velocity of light measured by natural clocks and rods is constant not only within the system, but that it receives the same numerical value in every inertial system. This statement contains axioms I to VIII and X, and definitions 1 to 21. It is not directly confirmed since axiom X has not been confirmed. The Michelson experiment is not relevant for this wider assertion. But the assertion becomes plausible on the assumption that there is no preferential inertial system. This most far-reaching conception underlies all of Einstein's presentations.

In all three interpretations the principle is a combination of axioms and definitions. Among the latter the definition of simultaneity is the most important. The definitions are neither true nor false but arbitrary; the principle is therefore neither a necessary consequence of the facts nor an incorrect or logically inadmissible one. On the other hand, the principle is not completely arbitrary in any of the three interpretations. Certain facts, the axioms, must be satisfied in order to make it possible to define time in such a way that the velocity of light will be constant. The existence of these facts can be determined experimentally in the same way as the existence of any other fact of physics.

§ 22. *Absolute Time*

By absolute time one understands the choice of a specific time metric (uniformity) as well as the choice of a specific simultaneity. The special theory of relativity is not concerned with the metric of time, only with simultaneity.

The thesis of absolute time admits three different interpretations. First, it might say that only a specific definition of simultaneity will lead to a unique description of all physical events. A simple epistemological consideration shows the error of such a conception (cf. § 2). Second, it might say, "It is possible to define simultaneity for every system in such a way that two events, simultaneous in an arbitrary system, are simultaneous in all other systems." This assertion is trivial and would not deserve the name "absolute" time. It is not disputed by the theory of relativity; any coordinate system may be selected and its time be called "absolute." Then one would have such a time. All other definitions of time would then be "pseudodefinitions" or "useful fictions." It is possible to indicate for any arbitrarily moving spatial coordinate system a more or less complicated definition of time so that the transformation into the preferred system will be $t = t'$. Concretely this means that every clock is constantly adjusted in such a way that it is always synchronous with that clock of the preferred system which it is passing at the moment. Such a definition of time would not be false, but it must not be called "absolute" because it could be carried through for every choice of simultaneity.

Only a third interpretation of absolute simultaneity is meaningful; it is the assertion, "There is a rule which if carried through in the same way in every coordinate system will always determine the same events as being simultaneous." Such a rule might employ an infinitely high velocity or the transport of clocks. It is essential in this context that the respective operation be the same for every system. It is an empirical question whether this assertion is true. In this sense, the determination of absolute time is an empirical matter. We shall investigate this question in the following sections.

It cannot be disputed, however, that certain definitions of time have logical advantages compared to others. It is an advantage if simultaneity becomes transitive, that is, if on the basis of the same operation of adjusting three clocks, *A* will be synchronous with *C* upon *A* being synchronous with *B* and *B* being synchronous with *C*. The addition that the operation be the same is necessary, because if the simultaneity between *A* and *C* is achieved in a different way (say, by the choice of different velocities of light for the trips to and from *C*), transitivity can always be achieved. It is another advantage if all causal chains become positive in time for a certain definition of simultaneity. But such rules do not necessarily lead to a single definition of simultaneity; they may leave a choice among different definitions. In addition, there is another indeterminateness because the rules themselves can be chosen arbitrarily. Yet the attempt of some philosophers at singling out one particular simultaneity in terms of an a priori intuition must be regarded as a failure, since any definition of simultaneity is just as intuitive as any other.

§ 23. *Absolute Transport Time*[32]†

The first possibility of defining the synchronism of clocks so that the use of the same rule will establish the same simultaneity in all systems is given by the transport of clocks. Two natural clocks synchronized at the same place are called synchronous if one of them is transported to a different place. This is a definition of simultaneity by means of the same rule for all systems as required by the above interpretation of absolute time; this definition uses a metrical method for the characterization of absolute simultaneity. It cannot be false because as a

[32] A report about the results of this section appeared in *Zeitschr. f. Physik*, vol. 9 [1922], p. 111. It also contains (p. 116 n.) a reference to F. Adler who makes use of transport time.

† See editor's reference notes, p. 204.

definition it is arbitrary. But if it is to be unique—and this requirement must be made for absolute simultaneity—the following axiom must be satisfied.

Axiom A. Two natural clocks that are synchronous at the same place are likewise synchronous if compared at the same place after being transported along different paths.

Whether axiom *A* is satisfied is a purely empirical question which can be answered independently of the definition of simultaneity for distant points and independently of the theory of relativity. It is denied by the theory of relativity, but since axiom *A* could not yet be tested directly, the question arises: From which axioms of the theory of relativity does the falsehood of axiom *A* follow?

We should like to simplify the problem in order to be able to treat it as a problem of the special theory of relativity. Included in axiom *A* is an accelerated motion of the clocks or a transport through gravitational fields, and it may be possible that in such a case a retardation of clocks will occur, in contrast to a uniform motion in an inertial system. The retardation would be established if a red shift were experimentally confirmed. Nothing, however, would be determined with respect to a retardation of clocks moving uniformly (cf. § 42). For the special theory of relativity the synchronism can already be defined by means of a uniform transport motion of the clocks. Imagine two clocks at different points, U_1 and U_2, which are to be synchronized. A third clock U_3 is transported uniformly from U_1 to U_2 and synchronized with U_1 at the moment of passing it. This synchronism is transferred to U_2 at the moment when U_3 passes U_2. In this way an absolute time would be defined for the special theory of relativity; its uniqueness presupposes the following weaker axiom.

Axiom B. The resulting synchronism is independent of the degree and the direction of the uniform transport velocity.

One difficulty arises. So long as simultaneity is not defined for any coordinate system, uniform translation is not defined either. At least in one system simultaneity ought to be defined according to Einstein. Then uniform translation would be defined relative to this system and axiom *B* would be applicable. We shall use this method to define "uniform" in axiom *B*. This is not an axiom but a definition. It is not necessary that the resulting transport synchronism in this system become identical with that of Einstein. It need merely be unique. Only later shall we add the restrictive assumption (axiom *C*) that a system exists in which this identity holds. Even then one is not compelled to use the preferred system as the starting system for the definition of simultaneity.

We shall now investigate the question: Which of the axioms I to X contradict axiom *B*? It is clear that axiom *B* can only contradict the matter axioms, not the light axioms, since it does not state anything about light. We shall omit two of the matter axioms, namely, axioms IX and X, and show first that axiom *B* is compatible with the other axioms.

We shall demonstrate this result by deriving from the combined axioms internally consistent transformation formulas for two uniformly moving systems K and K'. We can add arbitrary definitions to this derivation, because the choice of definitions cannot influence the result; only the form of the transformation formulas depends on the definitions, not their consistency. We merely simplify the calculation by choosing suitable definitions.

The first, omitted axiom, axiom IX, states that the length ($= l_K^{K'}$) of a rod at rest in K measured in K' is equal to the length ($= l_{K'}^{K}$) of the same rod at rest in K' measured in K; the axiom contains the relativity requirement for lengths. The second, omitted axiom, axiom X, refers to the principle of the constancy of the velocity of light and contains the partial assertion of this principle that the constant c receives the same numerical value in all systems if measurements are made with the same measuring rods and clocks (cf. § 21).

Corresponding to these axioms we shall omit definitions 16 and 17 and replace them by the following.

Definition a. The transfer of the unit of length is achieved by means of the transport of a rigid rod.

Definition b. The transfer of the unit of time is achieved by means of the transport of a natural clock.

These are suitable definitions. We shall add a restriction to the derivation of the formulas. We add an axiom *C*. When we have proved that axioms *B* and *C* are compatible with axioms I–VII and IX, then axiom *B* alone will certainly be compatible with them; the proof will therefore not be restricted. In addition, axiom *C* has a more far-reaching justification. Axiom *B* alone does not guarantee the transitivity of the transport synchronism which axiom *A* alone would do. Since Einstein's synchronism is transitive (this follows from the light axioms), however, the transport synchronism is also transitive if axiom *C* is added to axiom *B*. One will hardly construct a theory of absolute time in which transitivity does not hold. Even in the old ether theory axiom *C* holds; it characterizes the system at rest in the ether.

Axiom C. There is one system K in which the transport synchronism is identical with Einstein's synchronism according to definition 8.

It can be shown that this identity does not hold for the other systems. In spite of this fact we define the synchronism in a different system K' according to Einstein (definition 8), that is, we do not use the transport synchronism; according to the above remark, we are permitted to do so. We are merely searching for transformation formulas which do not exclude the transport synchronism.

From axioms I–VIII, from definitions 1–15, 18–21, *a*, and *b*, and from axioms B and C, the following transformation formulas for the transition from the preferred system K to a system K' moving uniformly relative to K can be derived:

$$\left.\begin{aligned}
x &= vt' + \frac{c}{c'}x', & x' &= \frac{x - vt}{\dfrac{c}{c'} - \dfrac{v^2}{c \cdot c'}} \\
y &= \frac{\sqrt{c^2 - v^2}}{c'} \cdot y', & y' &= \frac{c'}{\sqrt{c^2 - v^2}} \cdot y \\
z &= \frac{\sqrt{c^2 - v^2}}{c'} \cdot z', & z' &= \frac{c'}{\sqrt{c^2 - v^2}} \cdot z \\
t &= t' + \frac{vx'}{c \cdot c'}, & t' &= \frac{t - \dfrac{v}{c^2}x}{1 - \dfrac{v^2}{c^2}}.
\end{aligned}\right\} \qquad (62)$$

c and c' are the velocities of light in K and K', respectively; since axiom X has been omitted, they receive two different numerical values. v is the velocity of K', measured in K. Formulas (62) are consistent and the proof is complete.

Calculation. Let the starting point A' of K' coincide with the starting point A of K at the time t_1; K' is moving with the velocity v, measured in K, along the x axis of K. A light signal is sent from A' in the direction of the positive x axis at the time

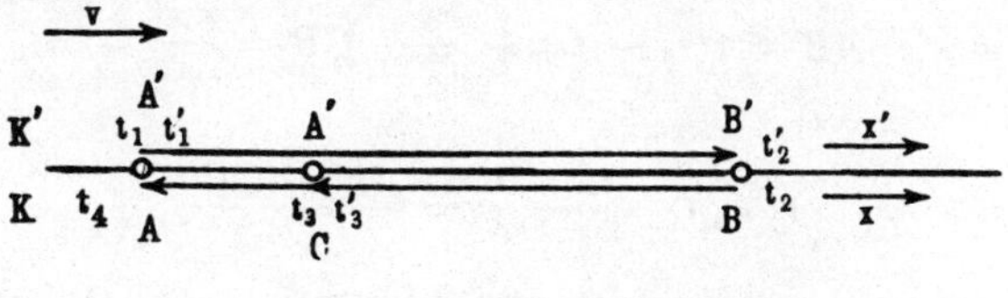

Fig. 11. Pertaining to the proof of formulas (62).

$t_1' = t_1$; it arrives at B at the time t_2' (in K: t_2) and returns to A' at the time t_3'. A' is now located above C of K; time in K: t_3. The signal continues and arrives at A at the time t_4 (fig. 11).

From axiom B and $t_1 = t'_1$, it follows that $t_3 = t'_3$. Furthermore, according to definition 8,

$$t'_2 = \frac{t'_1 + t'_3}{2} = \frac{t_1 + t_3}{2} \tag{63}$$

and likewise

$$t_2 = \frac{t_1 + t_4}{2} = \frac{t_1 + t_3}{2} + \frac{t_4 - t_3}{2}.$$

Then

$$t_4 - t_3 = \frac{AC}{c} = \frac{v(t_3 - t_1)}{c} = \frac{v(t'_3 - t'_1)}{c} = \frac{2vx'}{c \cdot c'},$$

because $t'_3 - t'_1 = \dfrac{2x'}{c'}$ as the time which the light in K' needed for the path $A'B'A' = 2x'$. Therefore,

$$t_2 = t'_2 + \frac{vx'}{cc'}. \tag{64}$$

This expression holds for any arbitrary point B′; if we omit the index "2," we obtain the time transformation in (62).

We shall now determine the linear transformation of the x axis. We call κ the contraction factor of the lengths; therefore, $l^K_{K'} = \kappa \cdot l^{K'}_K$ (same notation as in § 15). Because of the uniformity of the motion, the transformation is linear and, therefore, such a relation exists between the lengths. Thus

$$A'B' = x' \quad AB = v(t_2 - t_1) + \frac{\kappa x'}{c} \quad CB = \frac{\kappa x'}{c} - v(t_3 - t_2)$$

$$t_3 - t_1 = \frac{2A'B'}{c'} = \frac{AB + CB}{c}$$

By means of (63) and (64) and substitution in the last equation we obtain

$$\kappa = \frac{c}{c'} - \frac{v^2}{c \cdot c'}. \tag{65a}$$

The required equation has the form

$$x - vt = \kappa x' .$$

If (64) is substituted for t (the index "2" is omitted) and (65*a*) is substituted for κ, the first of equations (62) is obtained.

Finally, we determine the transformation perpendicular to the direction of motion. By analogy to theorem 20 we conclude that simultaneity perpendicular to the direction of motion means the same in K as in K'; we can draw this conclusion

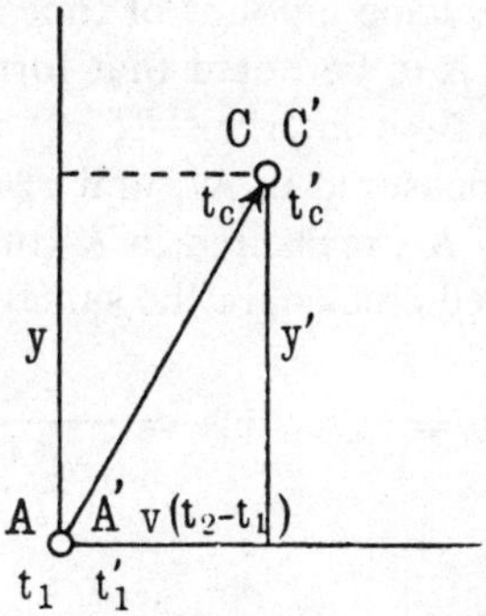

Fig. 12. Pertaining to equation (65*b*).

because the required transformation differs from the Lorentz transformation only in the choice of the units in K'. A signal is sent from A' at the time $t_1' = t_1$ to a point C' located on the y' axis; it arrives there at the time t_c'. According to theorem 20 and because of the fact that the clock at A' is always synchronous with that clock of K above which it is located at the moment (axiom B), $t_c' = t_c$; therefore the same result holds for the clock in C'. At this moment (because of theorem 20 the situation is the same in K as in K') the y' axis, seen from K, is parallel to the y axis (definition of parallelism and position of the axes in K'). Now

$$t_c' - t_1' = \frac{y'}{c'} \qquad (t_c - t_1)^2 = \frac{y^2}{c^2} + \frac{v^2(t_c - t_1)^2}{c^2} . \quad (65b)$$

From the second equation it follows that

$$(t_c - t_1)^2 = \frac{y^2}{c^2 - v^2}\,.$$

If $t_c - t_1 = t'_c - t'_1$, the second equation of (62) is obtained. The transformation for z is obtained by analogy.

The validity of axiom *B* can be recognized by the fact that according to the fourth equation on the left-hand side of (62), $t - t' = constant$ if $x' = constant$. A clock at rest in K' always differs by the same amount of time from the clocks of K which it passes. It is to be noted that formulas (62) exclude the relativity of the time intervals. If $\tau_K^{K'}$ is the period of a clock at rest in K, measured in K', and $\tau_{K'}^{K}$ the period of the same clock at rest in K', measured in K (that is, if the proper times of the compared clocks are the same), then

$$\tau_{K'}^{K} = \tau\,, \qquad \tau_K^{K'} = \frac{\tau}{1 - \dfrac{v^2}{c^2}}$$

and, therefore,

$$\tau_{K'}^{K} \neq \tau_K^{K'}\,. \tag{66}$$

With transformations (62) the principle of relativity for clocks is violated. But this result is plausible according to the assumptions of axioms *B* and *C*. The coordinate system determined by axiom *C* is a special one. This is the reason the inverse transformations in (62) have a form different from that of the direct transformations.

The question arises whether at least one of the omitted axioms can be added to this combination of axioms. We shall first add axiom X, that is, we shall set $c = c'$. Then equations (62) become

$$\left.\begin{aligned} x &= vt' + x'\,, & z &= \sqrt{1 - \frac{v^2}{c^2}}\cdot z' \\ y &= \sqrt{1 - \frac{v^2}{c^2}}\cdot y' & t &= t' + \frac{v}{c^2}\,x'\,, \end{aligned}\right\} \tag{67a}$$

$$\left.\begin{aligned} x' &= \frac{x - vt}{1 - \dfrac{v^2}{c^2}} & z' &= \frac{1}{\sqrt{1 - \dfrac{v^2}{c^2}}} \cdot z \\ y' &= \frac{1}{\sqrt{1 - \dfrac{v^2}{c^2}}} \cdot y & t' &= \frac{t - \dfrac{v^2}{c^2} x}{1 - \dfrac{v_2}{c^2}} . \end{aligned}\right\} \qquad (67b)$$

These formulas are also consistent. Therefore, axioms I to VIII, X, *B*, and *C* are compatible with one another.

Here the requirement of relativity is no longer satisfied for spatial distances either. From (67) the following expression is obtained for a distance *l* lying in the direction of motion:

$$l_K^{K'} = l , \qquad l_{K'}^{K} = l\left(1 - \frac{v^2}{c^2}\right)$$

and, therefore,

$$l_K^{K'} \neq l_{K'}^{K} .$$

This result is connected with the fact that (because of the validity of axiom VIII) the axes perpendicular to the direction of motion do not furnish identical transformations.

A different combination is obtained if axiom IX rather than axiom X is added to the above combination. From (62) the following result for a distance *l* lying in the direction of motion is obtained:

$$l_K^{K'} = \frac{c'}{c} \cdot l , \qquad l_{K'}^{K} = l\left(\frac{c}{c'} - \frac{v^2}{c \cdot c'}\right) ; \qquad (68)$$

with axiom IX, that is, with

$$l_K^{K'} = l_{K'}^{K} ,$$

the following condition results:

$$c'^2 = c^2 - v^2 . \qquad (69)$$

With this result axiom IX is also satisfied for the *y* and *z* direction, because now $y = y'$ and $z = z'$. This requirement must be made because of the application of theorem 21. If (69) is substituted in (62), the following expressions are obtained:

$$\left.\begin{aligned} x &= vt' + \frac{x'}{\sqrt{1-\frac{v^2}{c^2}}}, & x' &= \frac{x - vt}{\sqrt{1-\frac{v^2}{c^2}}} \\ y &= y' & y' &= y \\ z &= z' & z' &= z \\ t &= t' + \frac{vx}{c^2\sqrt{1-\frac{v^2}{c^2}}} & t' &= \frac{t - \frac{v^2}{c^2}x}{1-\frac{v^2}{c^2}}. \end{aligned}\right\} \quad (70)$$

Since (69) is a possible determination for c', (70) is consistent and axioms I to IX, *B*, and *C* are compatible with one another. The contradiction arises only when axiom X is added, for with $c = c'$ a contradiction arises from (69).

Let us summarize the result. The absolute transport time is excluded from the totality of the relativistic axioms, but this exclusion is based on axioms IX and X alone. The absolute transport time is even compatible with each individual axiom. Not even the principle of the constancy of the velocity of light alone can exclude the absolute transport time. If the velocity of light has the same numerical value for all systems, the absolute transport time is still possible. The Michelson experiment (axiom VIII) cannot exclude it. Only the combination of the "light principle" and the relativity principle makes the absolute transport time impossible. It is obvious how these two principles work together: axiom X merely states something about the relation between the clock unit and the unit of the measuring rod within the same system, and axiom IX determines the relation among units of measuring rods moving relative to one another. Only both axioms together determine the transformation of the units, and in such a way that the absolute transport time is excluded.

It will now be possible to compare the above formulas (62), (67), and (70) with the Lorentz transformation. We had shown in § 17 by means of what specific requirements the Lorentz transformation results from a transformation that leaves the spherical propagation of light invariant. Transformations (62), (67), and (70) are transformations compatible with the spherical propagation of light, because they contain all the earlier axioms and definitions referring to the propagation of light within one system. We must therefore be able to indicate the differences between these transformations and the Lorentz transformation with respect to requirements 2 to 6 of § 17.

If the formulas on the left-hand side of (62) are transferred into the symmetrical schema $x_1 \ldots x_4$ by means of the transformation

$$t = \frac{x_4}{c} \qquad t' = \frac{x_4'}{c'} \qquad v = v_0 \cdot c \tag{71}$$

and if these values are substituted in the expression $[x_v^2]$, then $k^2[x_v'^2]$ is obtained and, as can easily be calculated,

$$k^2 = \frac{c^2}{c'^2}(1 - v_0^2). \tag{72}$$

The difference from the Lorentz transformation consists solely in the modification of requirements 2 and 4 (§ 17), whereas all other requirements (also $v_2 = v_3 = 0$) are identical with those of the Lorentz transformation. The significance of axioms *B* and *C* consists in relation (72) which separates requirements 2 and 4 in a peculiar manner. Relation (72) is compatible with requirement 2 as well as with requirement 4, but not with both at once. If $c = c'$ in (72) according to requirement 4, then

$$k^2 = 1 - v_0^2 = 1 - \frac{v^2}{c^2}, \tag{73}$$

that is, $k^2 \neq 1$; this result corresponds to the transformations (67). If $k = 1$, then

$$c^2 - c^2 v_0^2 = c'^2,$$

that is,

$$c^2 - v^2 = c'^2$$

and, therefore, $c \neq c'$; this result corresponds to (69) and transformation (70). Only $c = c'$ and at the same time $k^2 = 1$ are not compatible with (72). Transformations (67) differ from the Lorentz transformation only by requirement 4, and transformation (70) only by requirement 2.

We are now in a position to formulate clearly the difference between relativistic time and absolute transport time. The transport time can also be defined light-geometrically; it is that light-geometrical time which furnishes the relation (72) for k when it is transformed. This definition makes sense even if no natural clocks and rigid rods are used. It can also be expressed intuitively corresponding to definition 16 without the use of definitions *a* and *b*. For this purpose relation (68) would be formulated in words and used as a definition; transformation (62) would be the result. It is then permissible to add one of the two definitions 16 or 17, but only one; in this case either transformation (70) or (67) will be obtained. The resulting light geometry is equivalent to the relativistic light geometry which differs from it merely by requirements 2 and 4, or rather by one of them. Another problem concerns the question: Which of these light geometries is identical with the geometry of rigid rods and clocks? The absolute transport time asserts in axioms *B* and *C* that this identity holds for the light geometry of relation (72), whereas the Einsteinian time asserts the same for the light geometry of requirements 2 and 4. This question whose answer depends on the validity of axioms IX and X can be decided only experimentally. If these axioms are true, the absolute transport time is false.

§ 24. *The Transverse Doppler Effect*

Axioms IX and X excluding the absolute transport time are just those axioms of the special theory of relativity that have not yet been confirmed experimentally. Therefore, an experimental test of axiom *B* is very desirable. It can be effected by means of an investigation of the transverse Doppler effect (for instance, by means of canal rays), because this effect expresses directly Einstein's retardation of clocks.†

The Doppler effect is the decrease (or increase) of the frequency of an oscillation which results when the emitting source is moving away from the observer (or is approaching him). Such a variation of frequency results for all kinds of periodic signals, for instance, sound signals. The amount of this frequency variation differs somewhat from that in classical kinematics, according to Einstein, because the retardation of clocks of the Lorentz transformation is superimposed upon the classical Doppler effect. But this difference is not measurable because it can always be ascribed to the degree of velocity of the source and this degree is not precisely measurable directly. However, a just measurable difference results if the source is moving transversally, that is perpendicularly to the line connecting the source and the observer. According to Einstein, a (very small) Doppler effect in the sense of a retardation of the frequency results even in this case, whereas it is zero in classical kinematics. This *transverse Doppler effect* is identical with the retardation of the Lorentz transformation.

If this effect should not occur, that is, if axiom *B* were confirmed, it would follow that either axiom IX or axiom X were false. Nothing definite would have been disproved concerning the light principle, because axiom X need not be false; either the relativity principle for rigid rods or the light principle would be false. Because of (66) the relativity principle for time intervals would be violated in any case. The transverse Doppler effect is to be viewed as an expression of the relativity principle rather than of the light principle.

Conversely, the question arises: What would follow for the validity of axioms IX and X if the experiment were to furnish the transverse Doppler effect to the required degree? This question can now be answered. Axiom *B* (and therefore also

---

† See editor's reference notes, p. 204,

axiom *C*) would be refuted, because the uniformly moving atomic clock would be retarded by the amount $\sqrt{1 - \frac{v^2}{c^2}}$.

In order to decide what would be derivable from such an occurrence, we shall look for the most general transformations compatible with axioms I to VIII and definitions 1 to 15, 18 to 21, *a*, and *b*. We obtain them from (62) by omitting axiom *B*, that is, by multiplying the units in *K'* by an arbitrary factor $1/\beta$. The new coefficients of measure x_i'' will result from the old ones by multiplication with β; therefore, $x_i'' = \beta x_i'$, $x_i' = \frac{x_i''}{\beta}$. If we write x_i' for x_i'', we obtain these most general transformations by substituting $\frac{x'}{\beta}\frac{y'}{\beta}\frac{z'}{\beta}\frac{t'}{\beta}$ for $x'y'z't'$ in (62); they become

$$\left.\begin{aligned}
x &= \frac{1}{\beta}\left(vt' + \frac{c}{c'}\,x'\right) & x' &= \frac{\beta(x - vt)}{\dfrac{c}{c'} - \dfrac{v^2}{cc'}} \\
y &= \frac{\sqrt{c^2 - v^2}}{\beta c'}\,y' & y' &= \frac{\beta c'}{\sqrt{c^2 - v^2}} \cdot y \\
z &= \frac{\sqrt{c^2 - v^2}}{\beta c'}\,z' & z' &= \frac{\beta c'}{\sqrt{c^2 - v^2}} \cdot z \\
t &= \frac{1}{\beta}\left(t' + \frac{vx'}{cc'}\right) & t' &= \frac{\beta\left(t - \dfrac{v}{c^2}\,x\right)}{1 - \dfrac{v^2}{c^2}}.
\end{aligned}\right\} \quad (74)$$

We add an axiom *D* expressing the state of affairs directly observed in the transverse Doppler effect.

Axiom D. A clock moving with the uniform velocity v relative to a certain system K experiences a retardation to the degree $\beta = \sqrt{1 - \frac{v^2}{c^2}}$.

At the outset, the "certain system K" would be the rest system of the laboratory in which the measurements are made; we shall show presently that this is not a privileged system, that is, that an axiom corresponding to axiom C does not occur.

From axioms I–VIII and D and from definitions 1–15, 18–21, a, and b the following transformations can be derived:

$$\left.\begin{aligned} x &= \frac{vt' + \dfrac{c}{c'}x'}{\sqrt{1-\dfrac{v^2}{c^2}}} & x' &= \frac{(x-vt)\cdot\sqrt{1-\dfrac{v^2}{c^2}}}{\dfrac{c}{c'}-\dfrac{v^2}{cc'}} \\ y &= \frac{c}{c'}\,y' & y' &= \frac{c'}{c}\,y \\ z &= \frac{c}{c'}\,z' & z' &= \frac{c'}{c}\,z \\ t &= \frac{t' + \dfrac{vx'}{cc'}}{\sqrt{1-\dfrac{v^2}{c^2}}} & t' &= \frac{t-\dfrac{v}{c^2}x}{\sqrt{1-\dfrac{v^2}{c^2}}}. \end{aligned}\right\} \qquad (75)$$

From this it follows that

$$\tau_{K'}^{K} = \frac{\tau}{\sqrt{1-\dfrac{v^2}{c^2}}} \qquad \tau_{K}^{K'} = \frac{\tau}{\sqrt{1-\dfrac{v^2}{c^2}}}$$

and, therefore,

$$\tau_{K'}^{K} = \tau_{K}^{K'},$$

that is, the retardation of axiom D holds conversely for a clock at rest in K, viewed from K'. Therefore, there is no preferred system K. It is easy to see that the amount of the retardation β can be derived from the requirement $\tau_{K'}^{K} = \tau_{K}^{K'}$; with $\tau_{K'}^{K} = \tau_{K}^{K'}$ the amount $\beta = \sqrt{1-\dfrac{v^2}{c^2}}$ can be derived from

(74). Thus axiom D is equivalent to axiom E which is obtained in the following manner.

Definition c. The time unit in K, measured in K', shall be equal to the time unit in K', measured in K.

Axiom E. The time unit transported by natural clocks in K' is equal to the time unit transferred according to definition c.

We shall now add one of the axioms IX and X to transformations (75). If we add axiom X, that is, if we set $c = c'$, the Lorentz transformation results. Then axiom IX is satisfied as a matter of course; therefore, axiom IX is a consequence of axioms I to VIII, X, and D (or E). If we add axiom IX, then

$$l_K^{K'} = \frac{c'}{c} \cdot \sqrt{1 - \frac{v^2}{c^2}} \qquad l_{K'}^{K} = \frac{1}{\sqrt{1 - \frac{v^2}{c^2}}} \cdot \left(\frac{c}{c'} - \frac{v^2}{cc'}\right),$$

and this result leads with $l_K^{K'} = l_{K'}^{K}$ to $c = c'$, that is, once more to the Lorentz transformation. (This result follows more simply from the transformation of the y coordinate which must result in $y = y'$ because of theorem 21. Axiom X is therefore a consequence of axioms I to IX, and D (or E).)[33]

Summarizing, we can say formulas (74) represent the most general transformation compatible with axioms I to VIII and definitions 1 to 15, 18 to 21, a, and b. $\beta = 1$ means the absolute transport time; $\beta = \sqrt{1 - \frac{v^2}{c^2}}$ means the Einsteinian retardation of clocks. If two of the axioms IX, X, and E (the latter may be replaced by D) are added to axioms I to VIII, the third axiom follows; on this condition, formulas (74) will represent a transition to the Lorentz transformation.

[33] In our construction of the axioms of the special theory of relativity we could have replaced definition 17 by definition c, and axiom X by axiom E; this would have been advantageous because of the symmetry to axiom IX. Nevertheless, we chose the given presentation because axiom X formulates Einstein's light principle.

If the transverse Doppler effect were quantitatively confirmed by experiment, formulas (75) would be confirmed empirically. Under these circumstances only one of the axioms of the Lorentz transformation would remain unconfirmed. Either axiom IX or X may be chosen as the unconfirmed axiom, since one of them will always become a consequence of the other.

If D (or E) were confirmed, but axioms IX and X disconfirmed (if one of them were false, the other one would be false), transformations (75) would replace the Lorentz transformation. They differ from it merely by the fact that, although the velocity of light is a constant for every system, it has a different numerical value in the various systems; in such a case the deviation would have to be ascribed to the measuring rods, not to the clocks.

§ 25. *Absolute Signal Time*

The second possibility of an absolute simultaneity consists in the attempt at defining synchronism at distant points by analogy with the neighborhood synchronism by a transition to a limit. In contrast with transport time, this attempt leads to a topological problem of time and is connected with axioms I and II. As in § 7, arbitrary signals are used which need not be light signals. They have the following properties, that is, they must satisfy the following axioms.

Axiom F,1. Given any two real points P and P′ and an arbitrary metric compatible with the time sequence at P, there exists at any time a signal PP′P so that $\overline{PP'P} < \delta$, *where* δ *is an arbitrarily given small magnitude.*

In other words, for first signals the time $\overline{PP'P} = 0$.

This axiom, like axiom *F,2*, is a topological axiom even though it makes use of a metric. But since the metric is arbitrary, the content of the axiom becomes independent of it; it is a topological condition that in the limit the time $\overline{PP'P} = 0$.

Axiom F,2. For every signal from P to P′ there exists a simultaneously departing signal reaching P′ earlier.

In other words, no first signal is a signal.

F,2 cannot be derived from *F,1*. With axiom I,1 it follows merely that the limit cannot itself be a signal in both directions, that is, that first signals exist which are not signals.

In addition, the signals must satisfy axioms I and II,4; from *F,1* alone it can then be derived that they satisfy axioms II,1–3, too. Therefore, theorems 1 to 6 are valid for these signals, and we shall use these theorems as well as axioms II,1–3 as derived theorems. The difference from the relativistic time of § 5 is represented by *F,1* alone which directly contradicts axiom II,5. Axiom *F,2* contradicts axiom III. As we shall show, axiom II,6 is also derivable for these signals.

We shall now define *absolute signal simultaneity.*

Definition d. If the first signal leaves P at the time t and arrives at P′ at the time t′, then t′ = t.

It follows from theorem 4 that a single time point at P' is characterized as simultaneous with t by this definition. We must now prove that, conversely, t is simultaneous with t', that is, that the defined simultaneity is symmetrical and transitive. These properties can be derived from axiom *F,1* alone.

Proof of symmetry. Let tt' be the first signal PP', and $t'T$ the first signal $P'P$. We assume first that $T < t$. Then a signal $t't_1$ exists so that $T < t_1 < t$. According to theorem 4, the first signal arrives at t_1' from t_1 so that $t_1' < t'$. Then a signal t_1t_2' exists so that $t_1' < t_2' < t'$. On these conditions the path $t't_1t_2't'$ is a temporally closed curve, a result excluded by axiom I,1. The assumption that $T < t$ leads therefore to a contradiction. We shall now assume that $T > t$. We search for two time points t_2t_3 so that $t < t_2 < t_3 < T$. The signal $t_2t_2't_3$ according to axiom *F,1* will lead to $t_2' > t'$; otherwise, tt' would not be a first signal. There exists another signal

$t't_2't_3$ reaching P before the signal $t'T$. Therefore, $t'T$ is not a first signal and this fact contradicts our assumption. The assumption that $T > t$ also leads to a contradiction and there remains only $T = t$; this means that the first signal tt' is also the first signal $t't$ and the simultaneity is symmetrical.

Proof of transitivity. Let tt' be the first signal PP', and $t't''$ the first signal $P'P''$. At first we assume that $t < t''$, that is, that the first signal PP'' leaving P at t will arrive at P'' at a time $t_0'' < t''$. Then a first signal $t'tt_0''$ exists which reaches P'' earlier than $t't''$; the latter is therefore not a first signal and this contradicts our assumption. We shall now assume that $t > t''$, that is, that a first signal $t''t_0$, $t_0 < t$ exists. Then, according to theorem 4, a signal $t''t_0t_0'$, $t_0' < t'$ exists, and $t''t'$ is not a first signal. Because of the demonstrated symmetry it follows from the assumption that $t't''$ as well as $t''t'$ are first signals. This is a contradiction. Therefore, t must be $= t''$, that is, the simultaneity is transitive.

It follows immediately that axiom II, 6 is valid; this assertion is satisfied in the absolute signal time, because generally for first signals $\overline{PP'P''\ldots P} = 0$.

Finally, we add axiom *F*,2 and prove the most important theorem about absolute signal time.

THEOREM α. If a time point t is given at P, there exists at P' one and only one time point t' so that neither a signal tt' nor a signal $t't$ exists. This time point t' is simultaneous with t.

Proof. It follows from definition *d*, from the symmetry of the simultaneity, and from axiom *F*, 2 that the time points $t = t'$ have this property. It can be shown as follows that there is no second point T' which has this property with respect to t. If $T' > t'$, there exists a signal tT' because t' is the lower limit of the signals PP' departing at t. If $T' < t'$, then the first signal which leaves P' at T' will arrive earlier than t at P according to theorem 4 and because $t't$ is also a first signal. Therefore, a signal $T't$ exists and there is no point other than t' with the properties of theorem α relative to t.

Theorem α cannot be proved by means of axiom *F,1* alone, without the help of axiom *F,2*. But axiom *F,2* serves merely to exclude the possibility that the lower limit of the signals be itself a signal. If this requirement is disregarded, the simultaneity retains its significance; above all, symmetry and transitivity hold, but there may be signals for which departure and arrival at different points will have to be called simultaneous. Expressed metrically, there are signals with infinite velocity (in one direction). This singularity could not be eliminated by any change in the definition of simultaneity.

Now, simultaneity at distant points has been uniquely defined for all systems by the same rule; it holds for all systems, because in axioms I, II, and *F* nothing is said about a selection of real points. Such a simultaneity may therefore be called absolute. There is no logical necessity for its use, but it possesses far-reaching advantages. It is that simultaneity in which all causal chains are in positive time; because of theorem α there is only a single definition of simultaneity which furnishes this result.

The difference from relativistic time is clearly expressed in axiom *F,1* which contradicts axiom II,5, in axiom *F,2* which contradicts axiom III, and in theorem α which contradicts theorem 8. These differences are topological, since they hold for every metric compatible with the time sequence. Even topologically, therefore, absolute signal time differs from relativistic signal time.

The decision on whether the relativistic or the absolute signal time holds depends solely on the validity of the axioms. Experience alone can furnish this decision, because experience alone can decide about the validity of the axioms. Is the time for first signals $\overline{PP'P} = 0$ or greater? The answer to this question must be given independently of any definition of simultaneity. We know that this time is greater than zero, because it is greater than zero for light signals and there are no faster signals (cf. § 20). Experience decides against absolute signal time.

§ 26. *Unreal Signals*†

At this point an objection can be made. Can signals faster than light not be produced by other means? Well-known examples are the point of intersection of two nearly parallel rulers one of which is moved transversally and the light ray of a rotating lamp which achieves a very high lateral velocity at a certain distance. It is easy to produce in such a way a velocity greater than that of light. Why are these "signals"—let us call them *unreal signals*—not suitable for a definition of simultaneity? Is it not possible to achieve an absolute signal time by means of them?

It should now be easy to answer this question, because we have merely to see whether the unreal signals satisfy the axioms of the absolute signal time. It is clear that axioms *F, 1* and *F, 2* are satisfied by them. But for the absolute signal time axiom I, 1 is also essential, because the existence of first signals depends on it. And this axiom is not satisfied by unreal signals.

This axiom can be satisfied only by signals which possess a definite direction. If there are two signals PP' and $PP''P'$ leaving P simultaneously and arriving at P' simultaneously, they can immediately be transformed into a closed curve in contradiction to axiom I, 1 if the direction of PP' may be interpreted in the opposite sense; the result is the path $PP''P'P$. It must not be objected that $PP''P'$ must also be reversed, because this requirement arises only *after* the definition of simultaneity has been given. The judgment whether axiom I, 1 is satisfied must be made independently of the definition of simultaneity by means of the marked direction of the signal. Only after the validity of axiom I,1 has been established in this manner can the simultaneity be defined consistently. If axiom I,1 is violated, theorem 3 is no longer valid. There is no lower limit at P' for the arrival of signals departing from P at the time t. If at P' any arbitrarily early time t' is given, there

† See editor's reference notes, p. 204.

exists a signal leaving P at the time t and arriving at P' at the time t'. It is not possible to discover by means of this signal whether it travels backward in time, because it has no mark indicating the direction.

It is precisely the peculiarity of the unreal signal that there is no mark by means of which it can be determined in which direction it travels. It is essential for causal chains that it can be decided by means of attached marks what is the cause and what is the effect; our definition of time order depends on this criterion. But there is no such mark for unreal signals and, therefore, they do not have a defined direction. It is impossible to decide in which direction the point of the intersection of the two rulers is moving. If a mark were attached to one end of the ruler, say, a projection of the edge, so that the point of intersection were to start later, nothing would change with respect to the arrival of the point of intersection at the other end. There is no mark characterizing the departure of the point of intersection at one end as the cause of its arrival at the other end. The moving point of intersection has a direction only after the simultaneity has already been defined. This direction depends upon the fact whether one of the rulers is slanted to the right or to the left relative to the other. Since one of the rulers is moving relative to the other one, its end points must be projected simultaneously into the system of the other ruler before it can be decided toward which side the ruler is slanted. If simultaneity is defined differently, the inclination of the moving ruler is also changed and the point of intersection runs in the opposite direction. There is no criterion according to which the opposite direction could be called false; the sense of direction can be determined only by a definition of simultaneity. Therefore, simultaneity cannot be defined conversely by an unreal signal.

We shall pursue the peculiarity of unreal signals in more detail by the example of the rotating light ray, because in this instance one hesitates to believe that the direction of the unreal signal can be conceived in the opposite sense. It can certainly be ascertained in which direction the lamp is rotated, because in order to answer this question we need only use genuine causal chains attaching directly to the lamp. Does not the same procedure indicate the direction of the rotating light ray on the outside? Can the light ray rotate in the opposite direction on the outside if it is produced by the lamp itself?

In order to answer these questions we should like to use a numerical example (fig. 13). Let a lamp emitting a narrow

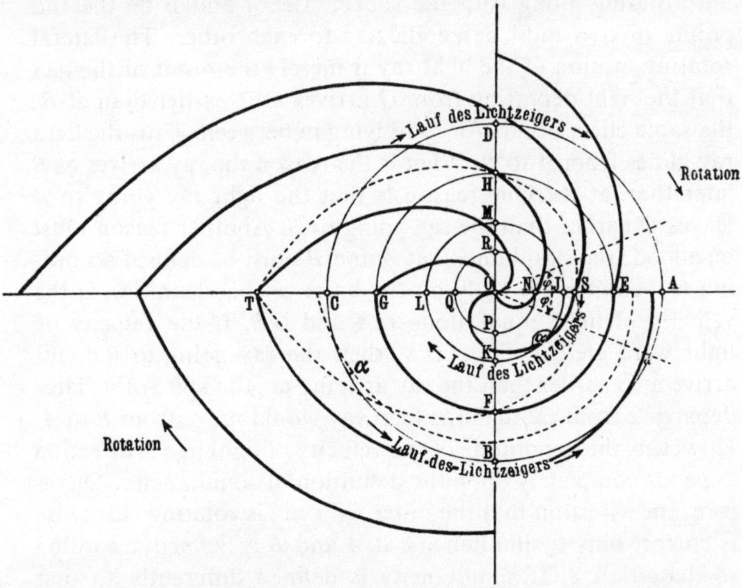

Fig. 13. Rotating light ray with definition (76) of simultaneity. Lauf des Lichtzeigers = Direction of light ray. Rotation = Rotation.

light ray be set up at O on a socket rotating clockwise at five hundred revolutions per second. (This number of revolutions lies only a little above the highest number technically achieved until now.) According to the customary calculation, the light ray travels already with the velocity of light on a circle with the radius $r = 95.5$ km (circle $NPQR$ of fig. 13). For larger circles it has a velocity greater than that of light and it might seem possible to use this signal for a time adjustment.

It seems as if it were objectively determinable that the light ray is moving clockwise. This statement is correct for the interior of the circle, $r = 95.5$ km, because a definition of time which here reverses the direction of rotation of the light ray would also reverse the direction of rotation of possible *real* signals. But how can we determine the direction of rotation outside of this circle? The light ray is not a rigid straight line rotating along with the socket. Let A and B be the end points of two radii perpendicular to each other. The lateral rotating motion of the light ray is merely the result of the fact that the light departing from O arrives at A earlier than at B; the same effect occurs for radii lying in between. Thus the light ray glides from A to B. What is the reason the ray arrives at B later than at A? One reason is that the light ray going to A leaves O earlier than the ray going to B. Another reason must be added: the simultaneity at A and B must be defined according to definition 8. Only on the basis of this definition is the velocity of light equal along OA and OB. If the velocity of light were greater along OB, then the ray going to B could arrive at B earlier than the ray arriving at A in spite of its later departure from O; then the light ray would move from B to A. However, the magnitude of the velocity of light in *one* direction depends completely upon the definition of simultaneity. Therefore, the assertion that the outer light ray is rotating clockwise is correct only if simultaneity at A and B is defined according to definition 8. If simultaneity is defined differently so that the ray toward A travels more slowly than toward B, then the direction of the light ray on the arc AB is reversed.

In figure 13 the relations for a different definition of time are indicated quantitatively. The simultaneity is adjusted according to

$$t_2 = t_1 + \frac{1}{2}(t_3 - t_1) + \frac{1}{2\pi}(t_3 - t_1)\left(\frac{\pi}{2} - \varphi\right). \qquad (76)$$

In the lower half of the figure from OA to $180° = \pi$, φ is to be counted in the direction of the rotation and in the upper half from OA to π in the opposite direction. Along OB and OD, (76) becomes $t_2 = \frac{1}{2}(t_1 + t_3)$, along OA it becomes $t_2 = t_1 + \frac{3}{4}(t_3 - t_1)$, along OC it becomes $t_2 = t_1 + \frac{1}{4}(t_3 - t_1)$.

(This choice of the definition of simultaneity does not contradict definition 2.) Along OB and OD the velocity of light $= c =$ 300,000 km/sec^{-1}, in the direction OA it is only $\frac{2}{3}c$, in the direction OC it is $2c$. If a nonrotating lamp is located at O, emitting light in all directions, then the propagation of light in the plane of drawing does not occur in circles as it does on the basis of the customary definition of time, but in heart-shaped curves bulging toward the left around O (the dotted curve $SFTH$ of fig. 13). These curves represent the position of the wave front at a time point of the time defined by (76). But if the *rotating* lamp with the *light ray* stands at O, then the ray has the form of the spirals drawn in figure 13; these spirals represent the position of the wave front at a certain time point. The spiral form results from the fact that the light is emitted later and later along the succeeding radii. All spirals have a bend at the horizontal axis which stems from the bend of the heart-shaped propagation curves.[34] (Spirals would also result for the customary time definition $t_2 = \frac{1}{2}(t_1 + t_3)$, but they would be circular spirals.) The spirals may now be imagined to rotate clockwise; during this process each spiral is gradually deformed into the next one. Let us consider the arc JKL in the lower half; it intersects the light ray at the obtuse angle α and, therefore, the light ray advances during the rotation in the direction JKL. Here the advance occurs in the direction of the rotation. On the semicircle EFG ($r = 300$ km) the light ray coincides with the arc. Here an infinite velocity exists; the whole wave front arrives simultaneously because of the faster advance along OB and the even faster one along OC. On the semicircle ABC the angle of intersection α is acute (in the figure an intermediate position of the light ray is indicated by a dotted line which is

---

[34] The bend results from the fact that in the upper half φ is counted in the opposite direction from that in which it is counted in the lower half. It can be avoided if φ is counted clockwise all around from OA and if, for instance,

$$t_2 = t_1 + \tfrac{1}{2}(t_3 - t_1) + \tfrac{1}{4}(t_3 - t_1)\cos\varphi\,.$$

Then a picture similar to figure 13 is obtained, but with continuously curved spirals. In this case the relations in the lower half of the figure become more complicated; the light ray travels on parts of the same semicircle in different directions.

likewise intersected at an acute angle α); here the wave front advances from C via B to A in the direction opposite to that of the rotation. Here a reversal of the direction of rotation has occurred. In the upper half of the figure, however, normal relations prevail. Whereas in the upper half the light ray travels on all semicircles in the direction of rotation, it travels in the lower half on all semicircles outside of EFG in the opposite direction.

It is obvious that the direction of rotation of the light ray can be reversed by means of a suitable definition of time. The resulting picture can be distinguished from the corresponding picture on the basis of the customary definition of time only by the deformation of the light rays; but this fact results in the reversal of the direction of rotation. This other direction of rotation of the light ray along CBA cannot be called wrong; there is no criterion by means of which it can be determined whether the light ray travels from C via B to A or from A via B to C. This is the reason the resulting velocity greater than that of light cannot be used for a definition of simultaneity. It is not a *real signal* satisfying axiom I,1.

The impossibility of defining simultaneity by means of an unreal signal can also be derived from another phenomenon connected with the reversibility. An infinite velocity is needed in order to produce absolute simultaneity. It lies on the semicircle EFG in figure 13; the arrival of the light at F and at G would therefore be called simultaneous with the arrival at E. This statement is correct for figure 13, but only because simultaneity was previously defined in this manner. If it is defined in a different way, then the infinitely fast signal lies on a completely different arc. It cannot be ascertained where the infinite velocity lies; this follows only after the definition of simultaneity has been given. Therefore, the use of an infinitely fast signal always leads only to that simultaneity which had previously been established by definition. It is different for *real* signals: for two *real* signals between P and P' it can be determined without the use of a definition of simultaneity which is the faster one. If there existed arbitrarily fast real signals, they would therefore define at P' a time point as a limit at which the infinitely fast signal arrives. For unreal signals it is in principle not decidable which is the faster one, and therefore they do not define a limit of the arrivals at P'. In figure 13 the unreal signal on the semicircle EFG is the fastest one, compared to the smaller as well as to the larger semicircles. On the basis of a different definition of time, however, EFG would not be the fastest signal; faster ones would exist on other semicircles. On the semicircle EFG other signals faster than those coming from the given lamp could be produced; for instance, a lamp with a different velocity of rotation could be established at O. On the basis of the definition of simultaneity (76), on the other hand, a change of the velocity of rotation of the lamp would result merely in a retardation of the light ray on EFG; even an increase in the velocity of rotation results in a retardation, in the travel direction GFE.

The fact that the infinite velocity does not occur as a *limit*, but can always be conceived of as a finite velocity, shows very clearly the unreal character of such a signal.

§ 27. *Two More Attempts at a Definition of Absolute Time*

The above two methods of obtaining a definition of absolute time are, of course, not the only ones; there are many other possibilities. We shall discuss two of them which their authors regarded as a refutation of Einstein's theory of time; they exhibit the characteristic mistakes in the defense of an absolute time. First, all such presentations overlook the fact that the new method achieves nothing but a definition of simultaneity, and that it does not disprove the *epistemological* relativity of simultaneity. It might disprove only the *physical* relativity of simultaneity and would then constitute an absolute time in the sense of the third interpretation of § 22, that is, in the sense of a uniform rule leading to the same simultaneity independently of the state of motion of the spatial coordinate systems. Furthermore, these authors overlook the fact that this interpretation leads from philosophical statements to physical ones. The uniqueness of the definition—which is, of course, required—depends on certain axioms whose validity is expressly denied by the theory of relativity. These authors, therefore, contradict empirical statements of the theory and must expect the decision not from conceptual advantages of their conception of time, but from the empirical facts.

The first of these attempts stems from F. Adler.[35] Originally, it did not concern simultaneity but the measurement of the length of a moving rod. Since such a measurement is equivalent to a definition of simultaneity, the attempt actually concerns this problem. Adler imagines two rods both of which have laterally protruding metal blades at their ends. If these rods are laid side by side, the blades would touch each other, that is, the rods are of equal length. It is assumed, in addition, that the rods consist of a nonconducting material. Let one of the rods be rigidly connected to a battery from which two wires run to the blades of the rod. Let the other rod be rigidly connected to a galvanometer from which two wires run to the blades of this rod. In such an arrangement the electric circuit is closed by the contact of the blades and the galvanometer will be deflected.

Let us now imagine the rods to be separated again. Let one rod, together with its corresponding apparatus, move uniformly in the direction of its own length. Let the other rod, together with its apparatus, be set up in such a fashion that the first one passes it. The blades will then touch each other on the left and on the right. Now the question arises: Do the blades touch each other simultaneously or not? Adler believes that an absolute answer can be given to this question. If the contact happens at both ends simultaneously, the electric circuit will be closed at this moment and the galvanometer will be deflected once. If the contact is not simultaneous, the electric circuit will not be closed at any time and the galvanometer will not be deflected. An empirical fact independent of the observer

[35] Friedrich Adler, *Ortszeit, Systemzeit, Zonenzeit* (Vienna: Vorwärtsverlag, 1920), p. 81.

will therefore decide about the simultaneity; every observer, in any given state of motion, must be able to determine whether or not the galvanometer connected to the rod is deflected.

If this train of thought were correct, the so-defined simultaneity could be called absolute in the sense of § 22; it would be defined by one rule for all states of motion. Adler pursues this idea further by using this arrangement for a comparison of lengths; he calls two moving rods equally long if the galvanometer is deflected. But this idea is merely the consequent continuation of the definition of simultaneity and we shall therefore restrict ourselves to this definition.

It is not an objection to the principle of this idea that the arrangement represents only a thought experiment whose actual performance would raise insurmountable technical difficulties because of the momentary contacts. But we must consider the fundamental complications entailed by the momentary contacts. These complications consist in the fact that the merely momentarily closed circuit may not be regarded as stationary. It is true for a stationary circuit that the whole circuit is without current so long as it is interrupted at one place. Yet the conditions are essentially different for a momentary starting of the current. It depends solely on the state of the immediate environment whether the current begins to flow upon contact with the blade at the left end. It is irrelevant for this question whether or not the blade at the right end makes a contact. Upon contact with the left blade a capacity current will enter the wire and travel to the galvanometer. No matter whether the right blade is touched or not, the galvanometer will be deflected; it will be deflected somewhat later, after the current has reached it. If the right blade makes contact, too, a little later than the left blade, another capacity current will originate, travel to the galvanometer, and increase the deflection. The deflection represents the sum of the two capacity currents. If the contacts on the right and on the left occur simultaneously, the process is exactly the same; only capacity currents occur and the deflection is just as great. It must not be inferred from the deflection of the galvanometer that the contacts of the blades were simultaneous. The deflection occurs also when these contacts are not simultaneous.

The processes at the two ends have nothing to do with each other as long as the effect has not progressed from one end to the other. But this action spreads only with the velocity of light, not with infinite velocity. Only if the time difference between the contacts on the left and on the right is so great that a light signal can travel from the left end to the right end will a disturbance occur and the deflection of the galvanometer be different. Yet the time order of two events that can be connected by a light signal remains unaffected by the time transformation of the special theory of relativity. The special theory of relativity is, therefore, in complete agreement with the thought experiment. If a deflection of a certain degree occurs, as it must if the blades make simultaneous contact, the special theory of relativity may just as well regard this deflection as produced by non-simultaneous contacts if a sufficiently fast sequence of events is assumed.

We can now see why the electric method fails to determine an absolute simultaneity. It is not unique; it determines the time order of the two contacts only within a certain interval and leaves it open which contact is to be called the earlier and which the later one. This indeterminateness stems from the fact that the transfer of action has an upper velocity limit. The above method is evidently nothing but a special attempt at carrying through an absolute signal time. It is tied to the axioms of this time, and the interval left open for the time order is the interval of theorem 8 which shrinks to a point only with the axioms of § 25 (theorem α). Therefore, the electric method of determining simultaneity by means of the absolute signal time must fail.[36]

---

[36] Adler's idea has since invaded the antirelativistic literature. Especially ingenious Einstein adversaries have replaced the galvanometer by a powder charge with an invitation to the representatives of relativistic simultaneity to sit down on the powder charge, since they are maintaining that the two contacts are not simultaneous. The representatives of the theory of relativity regret to decline this invitation since the powder charge—with a sufficiently sensitive ignition—will explode even if only one contact is made.

The other attempt to be mentioned stems from K. Grunsky.[37] He uses a measurement of length for the definition of simultaneity. We imagine the length of a rod marked off on the x axis of a coordinate system K with the rod at rest in K. Subsequently, the rod is moved uniformly in the direction of the x axis, first backward and then forward. Let its rest system now be called K'. At the moment the right end point of the rod appears above the right mark on the x axis of K, a time t_1 is recorded; when the left end point appears over the left mark, a time t_2 is recorded. Then $t_1 = t_2$, that is, these events are simultaneous. This definition is equivalent to saying that the length of a moving rod is equal to its length at rest; it requires $l_{K'}^{K} = l_{K}^{K}$ (same notation as in § 15). Grunsky does not notice that this is a definition; he believes that this result is necessary and uses it for the discovery of simultaneity. Once more this assumption commits the mistake of regarding a simultaneity with certain advantages as an absolute one. Its advantage would consist in simplifying the measurement of length; but this simplicity is obviously descriptive.

Apart from these considerations, it must be required that the definition be unique. It is unique for the two systems K and K'; two events simultaneous in K are also simultaneous in K' if they are tested according to the same rule. This is easily seen because the same process furnishes the simultaneity in K and K'. But this definition is not unique for arbitrary systems. Two events simultaneous in K, if a system K' with the velocity v is used, are not simultaneous in K if the determination is made by means of a system K'' with the velocity w. This result is a consequence of the Lorentz contraction. If two rods with equal rest lengths but with different velocities relative to K are marked off on the x axis by means of the simultaneity holding in K, they will furnish different lengths; if they are to furnish the same lengths, the definition of simultaneity must be different for both. Grunsky's definition of simultaneity violates axiom IX which states that rigid rods in motion adjust in a manner incompatible with the uniqueness of this definition.

[37] Philosophic dissertation (Tübingen, 1923).

This definition can also be distinguished from Einstein's theory of time by means of an experimentally testable assertion. Although axiom IX has not yet been confirmed directly, it is not advantageous for Grunsky's definition to doubt the truth of this axiom. If this axiom were replaced by a different one making Grunsky's definition of simultaneity along the x axis unique, this definition would not hold for rods lying perpendicular to the direction of motion in consequence of axiom VIII—which has been confirmed by the Michelson experiment and is not questioned. The simplicity gained for the x axis is lost for the y and z axes; in Einstein's theory it is just the other way around, since for the Lorentz transformation $l^K_{K'} = l^K_K$ along the y and z axes. Grunsky's definition cannot therefore be carried through in general and does not even lead to an absolute time in the sense of the third interpretation of § 22.

As a last way out there remains only a change of definition 21 of the length of a rigid rod. But this method would entail the introduction of *metrical forces* in the sense of § 18 and lead to the perfectly empty absolutism described as the second interpretation in § 22. With this result Grunsky's attempt loses any value.

§ 28. *Definition of Simultaneity by Means of a Velocity Less than That of Light*

It has often been objected to Einstein's theory of time that it could just as well be carried through by means of a velocity less than that of light, for instance, by means of the speed of sound. From the viewpoint of the general theory of relativity this method is admissible; but such a definition of time does not have the advantages of Einstein's definition. The difference can easily be formulated. Axioms I to V hold for sound if the systems are moving with a velocity less than that of sound and signals other than sound signals are excluded; in such systems a sound geometry can be carried through. It fails, however, as soon as faster-moving systems are used, since sound does not have the special property of light of being the fastest signal. Furthermore, the matter axioms VIII to X would not hold in the more slowly moving systems (except one), that is, the sound geometry would not be identical with the geometry of rods and clocks. In addition, the simultaneity obtained by definition 8 would at some points be of such a kind that certain causal processes, for instance, light signals or very fast traveling projectiles, would be assigned negative time intervals.

§ 29. *Consistency, Minimal Character, and Independence of the Axioms*

We can treat the problem of the consistency of axioms and definitions in common. Since they represent a system of theorems in four-dimensional geometry, they are consistent in the same sense as this geometry. Above all, the axioms alone, without the use of the definitions, are consistent. Thus the question of consistency has been clarified sufficiently for physics; the axioms contain at least possible facts. As to whether they actually occur or not, only experiments can tell.

It must, furthermore, be required that the axioms form a minimal system. This requirement is satisfied to a certain degree, but it is restricted by the requirement that the axioms are to contain elementary facts (§ 1).

Another essential requirement is mutual independence. This requirement is partly satisfied by the fact that the axioms refer to separate objects; for instance, the light axioms and the matter axioms are logically independent of each other. §§ 23 to 25 contribute to the discussion of independence; in these sections certain axioms are omitted and replaced by contradictory ones without the appearance of contradiction in the new system. Finally, the fourth part will furnish a contribution to this question by omitting more special axioms and by constructing a system by means of the other axioms. On the other hand, aside from certain exceptions, it will not be possible to omit general axioms and to construct systems by means of special axioms; the special axioms are frequently to be conceived of merely as additions to the general ones (cf. note on p. 89).

It would, however, be interesting to pursue the question of the minimal character and of the independence in greater detail.

Second Part

General Theory of Relativity

Chapter Three

The Axioms and the Construction of the Metric

§ 30. *Preliminary Mathematical Remarks Concerning the Metric in Infinitesimal Regions*

The Euclidean (more precisely, the pseudo-Euclidean) geometry of the four-dimensional, space-time continuum is characterized by the fact that in it the $g_{\mu\nu}$ adopt the normal form

$$\begin{matrix} 1 & 0 & 0 & 0 \\ 0 & 1 & 0 & 0 \\ 0 & 0 & 1 & 0 \\ 0 & 0 & 0 & -1 \end{matrix} = \overline{g_{\mu\nu}} \tag{77}$$

But if it is asserted that every geometry is Euclidean in the infinitesimal, an important distinction must be added. In the infinitesimal a coordinate system can be characterized in two ways:

1) by the requirement that the $g_{\mu\nu}$ are to adopt the normal form $\overline{g_{\mu\nu}}$;

2) by the requirement that the $\dfrac{\partial g_{\mu\nu}}{\partial x_\sigma}$ are to disappear.

The first requirement is easily satisfied. At a certain point the normal form can always be obtained by means of a linear transformation. If the values g_{ik} for the world point Q are given in an arbitrary coordinate system,[38] then the sixteen

[38] We shall always call world points Q and space points P.

constant coefficients a^i_μ for the transformation

$$x_i = a^i_\mu x_\mu \tag{78}$$

can be chosen in such a way that the ten equations are satisfied:

$$a^i_\mu a^k_\nu g_{ik} = \overline{g_{\mu\nu}}. \tag{79}$$

These equations contain only numbers, since only the numerical values of the g_{ik} occur at the point Q; their functional form is irrelevant. Outside the point Q, the $g'_{\mu\nu}$ do not have the normal values; only at Q is $g'_{\mu\nu} = \overline{g_{\mu\nu}}$. The solution of equations (79) merely presupposes that the $g_{\mu\nu}$ furnish an indefinite quadratic form of the inertial index 1 (cf. conditions in § 48).

Requirement 2, however, demands much more than requirement 1. The establishment of a system K^* characterized by requirement 2 is generally not possible by means of a linear transformation, because the values of the $g_{\mu\nu}$ at points in the neighborhood of Q must be considered. On the other hand, the first derivatives $\dfrac{\partial x_i}{\partial x^*_k}$ can be chosen arbitrarily with respect to this transformation, because the forty equations

$$\frac{\partial g^*_{\mu\nu}}{\partial x^*_\sigma} = 0 \tag{80}$$

can be satisfied by the choice of the forty second derivatives $\dfrac{\partial^2 x_i}{\partial x^*_\mu \partial x^*_\sigma}$ alone.

The transformation of the derivatives of the $g_{\mu\nu}$ is given by

$$\frac{\partial g'_{\mu\nu}}{\partial x'_\sigma} = \frac{\partial x_i}{\partial x'_\mu}\frac{dx_k}{\partial x'_\nu}\frac{\partial x_l}{\partial x'_\sigma}\cdot\frac{\partial g_{ik}}{\partial x_l} + \frac{\partial^2 x_i}{\partial x'_\mu \partial x'_\sigma}\cdot\frac{\partial x_k}{\partial x'_\nu}\,g_{ik} + \frac{\partial^2 x_i}{\partial x'_\nu \partial x'_\sigma}\cdot\frac{\partial x_k}{\partial x'_\mu}\,g_{ik}. \tag{81}$$

If these derivatives are to vanish in the special system K^*, then the right side of (81) must vanish if x'_ρ is replaced everywhere by x^*_ρ. From (81) forty equations are obtained which may be conceived of as pure numerical equations, because only the numerical values of the $\frac{\partial x_i}{\partial x^*_\mu} \frac{\partial^2 x_i}{\partial x^*_\mu \partial x^*_\sigma}$ enter at the point Q; no integrability conditions occur. These equations can be satisfied by the choice of the forty numerical values $\frac{\partial^2 x_i}{\partial x^*_\mu \partial x^*_\sigma}$ alone if the $\frac{\partial x_i}{\partial x^*_\mu}$ have arbitrarily given values.

If a system K^* that obeys requirement 2 is obtained, it is always possible to go from it to a system K_0 in which both requirements are satisfied. A linear transformation merely needs to be applied to K^* whose coefficients are determined according to (79). The property of K^* that the $\frac{\partial g^*_{\mu\nu}}{\partial x^*_\sigma}$ vanish is not lost, because $\frac{\partial g_{\mu\nu}}{\partial x_\sigma}$ is a tensor for linear transformations.

For linear transformations, the $\frac{\partial^2 x_i}{\partial x'_\mu \partial x'_\sigma}$ vanish and in (81) only the first term remains on the right-hand side. This expression is identical with the tensor transformation. Therefore, $\frac{\partial g'_{\mu\nu}}{\partial x'_\sigma} = 0$, if $\frac{\partial g_{ik}}{\partial x_l} = 0$.

We shall call a system satisfying requirements 1 and 2 for Q a *local inertial system*, and a system satisfying only requirement 1 a *local pseudo-orthogonal system*. In place of the latter designation we shall also use the briefer term *local orthogonal system* where no confusion will result from this usage. A system obeying only requirement 2 is called (by Weyl) a *local geodesic system*.

An example of a local orthogonal system in the static field of the earth is one consisting of the length, the width, and the height of a room, measured in centimeters, the time unit of which system is equal to the time light needs to travel one centimeter in this room. Simultaneity is to be defined according to definition 8. The corresponding local inertial system would be represented by a free-falling elevator in which the four coordinates are determined in the same manner. An example of a system satisfying requirement 2 but not requirement 1 is obtained if oblique spatial coordinates are introduced in the free-falling elevator and simultaneity is defined differently, for instance, by a linear transformation $x'_4 = x_4 + ax_1$. In nonstationary fields the local orthogonal systems are restricted to four-dimensional infinitesimal domains; the $g_{\mu\nu}$ retain the normal values only for a short time dx_4.

§ 31. *The Conservation Transformations in Infinitesimal Regions*

In this section some auxiliary concepts will be developed which we shall use later.

1. *Conservation of simultaneity.* If one goes from a system K to a system K' by means of the transformation

$$x'_i = f_i(x_1 \ldots x_4) \tag{82}$$

with the secondary restriction that for a certain point event Q with the coordinates $\overline{x_i}$

$$\frac{\partial x'_4}{\partial x_\alpha} = 0 \qquad \alpha = 1, 2, 3 \tag{83}$$

[39]

is to hold, then K and K' have the same simultaneity at Q. Point events in the neighborhood of Q which are simultaneous with Q in K have in K' the time coordinates

$$x'_4 = \overline{x'_4} + \frac{\partial x'_4}{\partial x_\alpha} dx_\alpha + \cdots \qquad \alpha = 1, 2, 3 .$$

($\alpha = 1, 2, 3$ since $dx_4 = 0$.) Because of (83) the terms of first order written above vanish so that the neighboring point events in K' are also simultaneous in first order. We speak therefore of the conservation of simultaneity in the infinitesimal. In this connection it is irrelevant how simultaneity is defined in K; only a relation between the two systems K and K' is relevant.

---

[39] If the index varies only from 1 to 3, it will always be called $\alpha, \beta, \gamma, \delta$. All other indices vary from 1 to 4.

Relation (83) is also valid for the inverse transformation. Thus

$$\frac{\partial x_4}{\partial x'_\alpha} = \frac{D_4^\alpha}{D} = 0 \qquad \alpha = 1, 2, 3, \tag{84}$$

where D is the determinant of the $\dfrac{\partial x'_\mu}{\partial x_\nu}$ and D_4^α the subdeterminant to $\dfrac{\partial x'_\alpha}{\partial x_4}$. These subdeterminants vanish, since in them a column $= 0$ because of (83). The infinitesimal conservation of simultaneity is therefore a symmetrical relation.

2. *Conservation of rest for a rod.* If one goes from K to K' by means of transformations (82) with the restriction for the world point Q

$$\frac{\partial x'_\alpha}{\partial x_4} = 0 \qquad \alpha = 1, 2, 3, \tag{85}$$

we speak of the conservation of rest for a rod.

We consider a spatial point $\bar{P}$ with the spatial coordinates $\overline{x_\alpha}$ at rest in K and coinciding at the time x_4 with a spatial point $\bar{P}'$ with the coordinates x'_α of the system K'. After the time dx_4, $\bar{P}$ in K has the same spatial coordinates $\overline{x_\alpha}$ (this is the definition of "rest" in the coordinate system); its coordinates x'_α in K' are

$$\begin{aligned} x'_\alpha &= \overline{x'_\alpha} + \frac{\partial x'_\alpha}{\partial x_\mu}\,dx_\mu + \cdots \qquad\qquad \alpha, \beta = 1, 2, 3, \\ &= \overline{x'_\alpha} + \frac{\partial x'_\alpha}{\partial x_\beta}\,dx_\beta + \frac{\partial x'_\alpha}{\partial x_4}\,dx_4 + \cdots \\ &= \overline{x'_\alpha} + \frac{\partial x'_\alpha}{\partial x_4}\,dx_4 + \cdots \end{aligned}$$

because of $dx_\beta = 0$. With (85) the term containing dx_4 also vanishes and in first order $x'_\alpha = \overline{x'_\alpha}$, that is, $\bar{P}$ is also at rest in K'.

Corresponding conditions hold for neighboring points. If P is a spatial point at rest in K with the spatial coordinates $\overline{x_\alpha} + dx_\alpha$, then it has in K' at the time $\overline{x_4}$ the spatial coordinates

$$\overline{x'_\alpha} + dx'_\alpha = \overline{x'_\alpha} + \frac{\partial x'_\alpha}{\partial x_\beta} dx_\beta + \cdots \tag{86}$$

After a time dx_4 it has in K' the coordinates

$$\overline{x'_\alpha} + \frac{\partial x'_\alpha}{\partial x_\beta} dx_\beta + \frac{\partial x'_\alpha}{dx_4} dx_4 + \cdots \tag{87}$$

The dx_β are equal in (86) and (87), because P is at rest in K. Its change of spatial position in K' is equal to the difference (86) – (87) and therefore with (85) in first order = 0.

Neighboring points, at rest in K, will therefore also be at rest in K' at the time $\overline{x'_4}$ in first order. If we imagine these two neighboring points in K to be connected by a rigid rod, it will also be possible to connect their positions in K' by a rigid rod at the time $\overline{x'_4}$. Strictly speaking, there exists only a mathematical relationship and we do not know yet whether or not a rigid rod can be at rest at all in K. But if an infinitesimal rigid rod can be at rest in K at the time $\overline{x_4}$, it will also be at rest in K' at this time. This result follows from the given relation and is the reason we speak of the conservation of rest for a rod.

Condition (85) holds also for the inverse transformation as can be concluded by analogy to requirement 1.

We add the following auxiliary concept.

Definition 22. A rigid rod is at rest differentially at a real point P at the time $\overline{x_4}$ if the coordinate system in which the rod is permanently at rest is connected to the system in which P is permanently at rest by a transformation which conserves rest for the rod.

At the time $\overline{x_4}$ the rigid rod has the same velocity as P but a different acceleration than P. In this case we speak of differential rest, because rest occurs only during a time element dx_4.

3. *Conservation of time order.* If one goes from K to K' by means of transformation (82) and the restriction for the world point Q

$$\frac{\partial x'_4}{\partial x_4} > 0, \tag{88}$$

we speak of conservation of time order. We consider two events occurring successively at a space point P in K; therefore, $dx_4 \neq 0$, $dx_\alpha = 0$. In K' their time interval becomes

$$dx'_4 = \frac{\partial x'_4}{\partial x_i} dx_i = \frac{\partial x'_4}{\partial x_\alpha} dx_\alpha + \frac{\partial x'_4}{\partial x_4} dx_4 + \cdots$$

Because of $dx_\alpha = 0$, the first term on the right-hand side vanishes. If (88) is satisfied, dx'_4 has the same sign as dx_4; thus the direction of temporal succession is preserved.

In general, relation (88) is not symmetrical and therefore $\dfrac{\partial x_4}{\partial x'_4}$ is not necessarily > 0. It will be symmetrical, however, if conservation of rest for a rod or conservation of simultaneity is added. Thus

$$\frac{\partial x_4}{\partial x'_4} = \frac{D^4_4}{D} = \frac{1}{\dfrac{\partial x'_4}{\partial x_4}} \quad \text{if} \quad D = \frac{\partial x'_4}{\partial x_4} \cdot D^4_4. \tag{89}$$

The latter is satisfied by (83) or (85).

4. *Conservation of uniformity.* If one goes from K to K' by means of transformation (82) and the restriction for the world point Q

$$\frac{\partial^2 x'_4}{\partial {x_4}^2} = 0, \tag{90}$$

we speak of conservation of uniformity at Q. For two point events occurring at the same place in K ($dx_\alpha = 0$, $\alpha = 1, 2, 3$)

and differing from each other by dx_4

$$dx'_4 = \frac{\partial x'_4}{\partial x_4} dx_4 + \frac{\partial^2 x'_4}{\partial x_4{}^2} dx_4{}^2 + \cdots \tag{91}$$

$\frac{\partial x'_4}{\partial x_4}$ is the ratio of the time units of the two systems; it may be arbitrary. The next term is the deviation of first order in the period of the clocks; (90) means, therefore, that the periods of the clocks in K and K' are equal in first order.

Requirement (90) holds only with certain restrictions for the inverse transformation. Thus

$$\frac{\partial^2 x_4}{\partial x_4'^2} = \frac{\partial}{\partial x'_4}\left(\frac{\partial x_4}{\partial x'_4}\right) = \frac{\partial}{\partial x_\mu}\left(\frac{\partial x_4}{\partial x'_4}\right) \cdot \frac{\partial x_\mu}{\partial x'_4} = \frac{\partial}{\partial x_\mu}\left(\frac{D_4^4}{D}\right) \cdot \frac{\partial x_\mu}{\partial x'_4}$$

$$= \frac{1}{D^2} \cdot \left[D \cdot \frac{\partial D_4^4}{\partial x_\mu} - D_4^4 \frac{\partial D}{\partial x_\mu}\right] \frac{\partial x_\mu}{\partial x'_4} . \tag{92}$$

If this expression is to vanish, then with $D = D_4^\nu \frac{\partial x'_\nu}{\partial x_4}$ the following condition results:

$$\left[D_4^\nu \frac{\partial D_4^4}{\partial x_\mu} \frac{\partial x'_\nu}{\partial x_4} - D_4^4 \frac{\partial D_4^\nu}{\partial x_\mu} \frac{\partial x'_\nu}{\partial x_4} - D_4^4 \, D_4^\nu \frac{\partial^2 x'_\nu}{\partial x_\mu \partial x_4}\right] \frac{\partial x_\mu}{\partial x'_4} = 0 . \tag{93}$$

Requirement (90) has merely the effect that in the third term the summand for $\mu = \nu = 4$ vanishes, whereas nothing is prescribed for any other term. In general, this expression will not vanish; therefore, $\frac{\partial^2 x_4}{\partial x_4'^2} \neq 0$, although $\frac{\partial^2 x'_4}{\partial x_4{}^2} = 0$. However, two other cases occur for which (93) vanishes:

a) For linear transformations. The second derivatives vanish and as a consequence also $\frac{\partial D_4^\nu}{\partial x_\mu}$, that is, every individual summand vanishes.

b) For those transformations in which in addition to the conservation of uniformity, conservation of simultaneity and of rest for a rod holds. Because of (85) in the first two terms ν must be $= 4$, since otherwise $\frac{\partial x_\nu}{\partial x'_4} = 0$, and these two terms will cancel each other out. In the third term we must also set $\nu = 4$, since $D_4^\nu = D \cdot \frac{\partial x_4}{\partial x'_\nu}$ vanishes because of (83) for $\nu = 1, 2, 3$; likewise $\mu = 4$, since otherwise $\frac{\partial x_\mu}{\partial x'_4}$ vanishes. Because of (90) the third term vanishes, too.

Aside from the trivial case of the linear transformation[40] and certain special cases,[41] the conservation of uniformity is therefore a symmetrical relation only if it is combined with the conservation of simultaneity and of rest for a rod.

§ 32. *Content of the General Theory*

The assertions of the general theory of relativity can be divided into three groups:

1. The assertion that a space-time metric with certain properties can be constructed by means of light, rigid rods, and natural clocks.
2. The assertion that all natural laws are covariant relative to this metric.
3. The assertion that the coefficients $g_{\mu\nu}$ of this metric lead to a theory of gravitation.

Only the analysis of the first of these assertions is an object of the present investigation; this is the reason the other two are not further discussed in the following sections.

The second assertion is usually called the relativity principle; it is an empirically testable assertion, because the covariance of the natural laws can no longer be established arbitrarily after the metric has been stipulated. This assertion is formulated axiomatically and cannot be analyzed any further.

---

[40] In this case, (83) and (85) need *not* be satisfied. For the Lorentz transformation only conservation of uniformity holds, not that of simultaneity and rest for a rod.

[41] Special numerical values can always be given to the first and second derivatives so that (93) will be satisfied.

The third assertion contains the equivalence principle; but it is even more general. If it were axiomatized, Einstein's field equations of gravitation, the motion of a mass point, etc., would result. This assertion is also empirically testable.

The physical significance of the theory of relativity consists especially in the second and third assertion. But a clarification of the first assertion is of interest to the two others, because it constitutes their logical and intuitive foundations. It is not obvious what part of the metric is empirical and what part is logical construction. The answer to this question is essentially given by the axiomatization of the special theory of relativity. With respect to the general theory, it will be necessary to show the revisions that must be made in the metric of the special theory.

§ 33. *The Space-Time Axioms of the General Theory*

We start with the following axioms.

Axiom XI,1. Axioms I to X hold approximately in any small region for any real point P as starting point, and for such points only. This approximation increases without limit as the region decreases.

Axiom XI,2 Clocks and measuring rods permanently at rest at a real point P behave in the same way as clocks and measuring rods differentially at rest at P.

Definition 23. A system of real points is called a spatial coordinate system if it is everywhere present (definition 5) and infinitesimally stationary at every point (definition 7).

Definition 24. For an arbitrarily chosen spatial coordinate system and an arbitrarily chosen adjustment of time, the metrical determination of the world is to be made in such a way that it will become identical at every point with the metrical determination that is locally prescribed by axioms I to X and definitions 1 to 21 while the same measuring rods and clocks are employed everywhere.

It should be noted how we obtain a definition of the spatial coordinate system: first, we define "everywhere present," then "stationary coordinate system," and finally "general coordinate system" as a generalization of the stationary one. The term "everywhere present" is not sufficient for a characterization of the coordinate system, because the requirement of continuity has to be added; this is achieved by requiring infinitesimal stationarity. The concept of coordinate system obtained in this manner is sufficiently general, because for any two arbitrarily close real points in any state of motion it can always be arranged that they belong to the same coordinate system. Nevertheless, not every system of everywhere present real points is a coordinate system. The character of the three-dimensional continuum, which is not yet given with continuity, is guaranteed by axiom V, which holds infinitesimally in the stationary coordinate system because of axiom XI, 1.

Axioms XI are differential axioms. One might assume that an integral axiom stating the existence of a coordinate system must be added. Such an assumption is probably necessary if the existence of a coordinate system for the world as a whole is required; it is not necessary, however, if only finite regions are concerned. It follows already from the validity of the special theory of relativity in the infinitesimal that around every world point a finite region can be defined in which a coordinate system according to definition 23 exists. Since we shall discover (§ 48) that we can assert integral properties in the general case only for such finite regions, we shall refrain from making a statement concerning the existence of a coordinate system for the world as a whole. A decision whether this more comprehensive assertion can be derived from the validity of the special theory of relativity in infinitesimal regions would be interesting.

The following questions must now be investigated:

1. It must be shown that axioms XI are actually contained in the general theory of relativity.

2. In must be investigated how definition 24 can be carried through and what minimum of measuring devices is required for it.

In this section the construction of the metric can be carried through only differentially, because only differential axioms are at our disposal.

§ 34. *The Validity of Axiom XI, 1*†

Axiom XI, 1 is characterized by the fact that no use is made of the existence of local inertial systems; it states that the special theory of relativity holds infinitesimally for every coordinate system formed by real points. It will be shown in the following way that this assertion is contained in the theory of relativity.

It suffices for the validity of axioms I to X in the infinitesimal that the local orthogonal system has been constructed; the additional requirement $\frac{\partial g_{\mu\nu}}{\partial x_\sigma} = 0$ is not necessary. The time dx_4 for a light signal $ds^2 = g_{\mu\nu} dx_\mu dx_\nu = 0$ will be

$$dx_4 = -\frac{d\gamma}{g_{44}} \pm \sqrt{\frac{d\gamma^2}{{g_{44}}^2} - \frac{d\sigma^2}{g_{44}}} \qquad \begin{aligned} d\gamma &= g_{\alpha 4}\, dx_\alpha \\ d\sigma^2 &= g_{\alpha\beta}\, dx_\alpha\, dx_\beta \\ \alpha, \beta &= 1, 2, 3\,. \end{aligned} \tag{94}$$

For the local orthogonal system this reduces to

$$dx_4 = d\sigma, \tag{95}$$

that is, the velocity of light becomes constant and $= 1$; the light axioms are valid. In the neighborhood of the location considered, the $g_{\mu\nu}$ can be represented by

$$\overline{g_{\mu\nu}} + \frac{\overline{\partial g_{\mu\nu}}}{\partial x_\sigma}\, dx_\sigma + \cdots \tag{96}$$

If this expression is introduced in (94), then it will also reduce to (95) for the limit $dx_\sigma = 0$. The validity of the light axioms in the limit $dx_\sigma = 0$ depends therefore only on the $g_{\mu\nu}$, not on their derivatives. But wherever the light axioms are valid the matter axioms are valid. Whatever is valid about them in

† See editor's reference notes, p. 204.

the local inertial system is also valid in the local orthogonal system, modified merely by terms of the order of magnitude dx_σ which vanish in the limit.

This means that a class of local systems exists in which axioms I to X are valid and which is much more comprehensive than the class of local inertial systems. This more general class can be made even more general. Axioms I to X are valid independently of the definition of simultaneity and the position of the spatial axes; they will therefore be valid in every local system that can be transformed into a local orthogonal system with conservation of time order, of rest for a rod, and of uniformity. We shall investigate such systems more thoroughly in § 46 and call them "real systems." It is implicit in the special theory of relativity to assume the existence of real points wherever axioms I to X are valid. No real points can exist outside this region, because they would have a velocity greater than that of light relative to other real points. Therefore, axioms I to X are infinitesimally valid for every real point as starting point, and only for such points.

The continuity and differentiability of the $g_{\mu\nu}$ are presuppositions of this conclusion; these presuppositions will always be retained in the following as being essential to the general theory of relativity. Singularities may only be "apparent"; a strict formulation will be given in § 46.

What is the significance of the local inertial systems? They have merely the advantage that axioms I to X hold in them with a higher approximation. Because of $\frac{\partial g_{\mu\nu}}{\partial x_\sigma} = 0$ the terms of first order disappear in (96) and a deviation from axioms I to X is produced only by the terms of second order.

§ 35. *The Validity of Axiom XI, 2*

We are going to show that axiom XI, 2 is also asserted by the general theory of relativity. It is still left open in axiom XI, 1 what state of motion the clocks and measuring rods must have if axioms I to X are to hold. However, they must—as stated in axioms I to X—be at rest during the measurement, that is, be at rest differentially at *P*. Three possible assumptions remain:

1. The clocks must be permanently at rest at *P*.

2. The clocks must be permanently at rest in certain special systems (e.g., inertial systems) and be at rest only differentially at *P*.

3. It is irrelevant in which system the clocks and measuring rods are at rest permanently if only they are at rest differentially at *P*.

Axiom XI, 2 asserts the third case; the second case might be possible, but the first case hardly comes into question.

All three assumptions have in common that the clocks and measuring rods are at least differentially at rest at *P*. Since axioms I to X hold for *P* according to axiom XI, 1, they are also at rest in a local orthogonal system obtained by a construction of the metric according to definitions 1 to 21. If one goes from this orthogonal system to the local inertial system, this transition can be achieved solely by the choice of the second derivatives of the transformation; an inertial system has, therefore, the same state of motion also as *P*.

At every real point—and only there—a rigid rod and a clock exist, since in axiom XI, 1 the matter axioms are mentioned. Every rod and every clock in any state of motion will therefore be at rest differentially in an inertial system. Accordingly, axiom XI, 2 asserts that every rigid rod l (and every clock) behaves in the same way as a rigid rod l_0 (or clock) which is permanently at rest in an inertial system and has momentarily the same velocity as, but a different acceleration from, the first.

This result is by no means obvious. It could happen that the acceleration deforms the rod, more precisely speaking, that the quotient l/l_0 will not approach 1 even for the limit $l = 0$. However, $\left(\frac{l}{l_0}\right)_{lim\ l=0} = 1$ is an assumption of the general theory of relativity. From this assumption follow, for instance, Einstein's well-known statements concerning the behavior of rods and clocks in a static gravitational field. It is also presupposed in the solution of the clock paradox. Axiom XI,2 is therefore asserted by the theory of relativity.

Here a difficulty arises. The rigid rod and the natural clock are defined as closed systems; but closed systems exist only in inertial systems. Although gravitation as a metrical force satisfying the properties of the force X mentioned in § 18 may be disregarded, every body at rest in the gravitational field is, in addition, exposed to a physical (e.g., elastic) force which prevents its falling motion and is not of the kind X. With what justification can one speak of rigid rods and natural clocks in the gravitational field?

This difficulty can be avoided if we restrict ourselves to the second of the three assumptions mentioned above and do not use the third one as does axiom XI,2. If free-falling material things are used, the physical supporting forces vanish if these material things are chosen sufficiently small so that they can be regarded as mass points on geodesic paths. Under these circumstances we may speak in the limit of infinitesimal, closed systems. Then rigid rods and natural clocks have their metrical significance only if they are permanently at rest in the same inertial system. It would be possible to restrict the theory to this assumption.[42] The theory still goes beyond the special theory of relativity; although the rigid rod is permanently at rest in the same inertial system, its world is distinguished from the special theory of relativity by the fact that its inertial system is not moving uniformly relative to the inertial systems of other world points. Another assumption would have to be added to assumption 2 stating that the changes under the influence of physical forces in the gravitational field are not permanent so that a transport of the units through noninertial systems might be possible. Such an additional assumption might say that assumption 2 (in place of axiom

---

[42] Occasionally, the second assumption is made with the addition that, aside from inertial systems, stationary systems $\left(\frac{\partial g_{\mu\nu}}{\partial x_4} = 0\right)$ are admissible; cf. F. Kottler, *Physikal. Zeitschr.*, vol. 22 (1921), p. 275. But then the problem of the closed system arises, as it does for the third assumption.

XI,2) is added to axiom XI,1 only during the measurement, whereas otherwise the state of the measuring rods is arbitrary.

The difficulty can also be eliminated from the viewpoint of assumption 3 and thus of axiom XI,2. It is possible to regard systems as approximately closed even in gravitational fields if the influence of the elastic counterforce is small relative to the other forces in the interior of the respective system. Thus a spring-balance clock in a gravitational field is not a closed system, because it experiences an elastic force by resting on a table; as a consequence an elastic bending of the gears and shafts and a disturbance of its period occur. Strictly speaking, a spring-balance clock in a gravitational field will run differently from a similar free-falling clock, and it will also run differently at points of different gravitational field strength. This effect can be neglected, because the elastic counterforces are small compared to the elastic forces which keep the spring balance in motion. However, this is no longer the case if the clock is fastened to a rotating disk and brought into strong centrifugal fields. Because of the elastic deformations it will no longer register the proper time.[43] This is also the reason a pendulum clock cannot be regarded as a natural clock. The supporting force of its hanging mechanism is not small compared to the inertia of its pendulum, but of the same order of magnitude, and in an inertial system the clock would not swing at all.

If we now add to the limiting process for a vanishing extension of material things, necessary for assumption 2, a second limiting process for vanishing physical counterforces, we can define "infinitesimally closed."

Definition 25. A material system is called infinitesimally closed if it can be chosen sufficiently small relative to space-time changes of the gravitational field strength and if the quotient of external physical supporting forces and internal physical forces approaches 0 with the decrease of the material system.

---

[43] This is the explanation of the problem mentioned by Mr. v. Laue. Cf. M. von Laue, *Relativitätstheorie II* (Friedr. Vieweg & Sohn, Braunschweig, 1921), p. 189.

This definition is satisfied by rigid rods whose elastic bending vanishes with a decrease of the rod relative to its internal stability. With respect to clocks this process cannot be accomplished by a decrease in the size of a particular clock; the definition means that smaller clocks are more perfectly closed systems than larger ones.

With the help of definition 25, axiom XI,2 is compatible with the concepts of rigid rod and natural clock.

In this context we should like to add a remark concerning the *atomic clock.* If statements are made about the frequency changes of this clock, for instance, when the red shift of the spectral lines of the sun is calculated, axiom XI,2 is usually presupposed. This assumption is not necessary; assumption 2 would suffice. The individual gas atoms are not bound elastically in the gravitational field so that they move on the geodesic line; the atom represents a clock permanently at rest in the same inertial system. Only at the moment of a coincidence with other atoms do external constraining forces occur; therefore, the assumption would have to be added that no permanent changes will occur. The emission of light will generally occur on free intervals of the path, since their duration is long compared to the duration of the coincidences.

In contrast, the individual electrons within the atom do not move on the geodesic line, since they are subject to electric forces. The entire atom, and not the electron alone, is to be regarded as a "clock"; the electric forces affecting the electrons are the internal driving forces of the clock. Two objections can be made to the conception of the atom as a clock. The first objection concerns the fact that the atom is not a closed system at the moment of radiation, because by way of the radiation it comes into contact with the environment. But the reactive effect of the emission upon the atom, that is, the recoil impulse and energy loss, is small compared to the internal electric forces of the atom so that the requirement of definition 25 remains satisfied. The second objection concerns the origin of the emitted frequency. According to classical theory, the frequency would be a direct measure of the period of the electron's revolution. But according to Bohr, the value of the frequency is a function of the energy loss when the electron jumps from one orbit to another. There is a connection between this frequency, however, and the frequency of the period. Yet, may we actually regard the atom as a clock if we accept Bohr's model? No definite answer can be given to this question. We can only say, if any mechanism has the properties of a clock according to Einstein, then the atom has them. The precision required by the theory of relativity cannot be demanded of ordinary clocks; their periods are not so accurately defined. At best, "the astronomical clock," that is, the planet rotating around its own axis, may be defined with sufficient precision. But a planet is not subjected to sufficiently great velocities and accelerations to furnish relativity effects. Therefore, the atomic clock will probably have to be regarded as the best clock, and one should adapt the definition of a clock to Bohr's ideas rather than reject the use of atoms as clocks.

No requirement corresponding to axiom XI,2 is made for light. The reason is that for light there is no rest system characterized by a real point and that it is impossible to influence the state of motion of light. The distinction between the three possible assumptions underlying axiom XI,2 is therefore irrelevant. Light always behaves according to the second assumption, because it moves on geodesic lines and thus always uniformly relative to the respective local inertial system (this assumption holds for light in place of rest for rods and clocks in the inertial system). But if axiom XI,1 asserts that the light axioms hold infinitesimally also for coordinate systems in any state of motion, it goes beyond the special theory of relativity. It goes beyond the limits of this theory to state something about the motion of light if the inertial systems of the various world points no longer combine into a rigid system. One could imagine, for instance, that light loses its special character in the gravitational field and no longer corresponds to the $ds^2 = 0$ characteristic of material media. Axiom XI,1, however, excludes this possibility.

§ 36. *Light Geometry in a Gravitational Field*

We shall now turn to the second of the questions mentioned in § 33. When we treat the construction of the metric from the axioms, we shall begin by investigating to what extent light signals alone suffice. We shall notice an essential difference between the special and the general theory of relativity in this connection. At the same time there will be an opportunity to give an additional proof connected with the special theory.

It has been shown in the first chapter (§ 16) that the light axioms I to V suffice by themselves to determine the coordinate system up to a linear transformation, and it has been stated that inertial systems are therefore determined by properties of light alone. Yet this assertion has not been sufficiently justified. In that chapter the uniqueness of the light geometry (up to a uniform translation) was demonstrated, but such a demonstration does not yet show that whenever the light axioms I to V are satisfied, the matter axioms VII to X are necessarily satisfied (§ 19). One might imagine a state of affairs in which the geometry of rigid rods and clocks becomes non-Euclidean in spite of the validity of axioms I to V. From the viewpoint of this geometry a gravitational field would exist which is of such a special structure that the light geometry becomes Euclidean. We wish to show that this cannot be the case; more precisely, we wish to show that Einstein's theory contains the assumption that this is not the case. Nothing else can be demonstrated here nor is anything else at issue.

Only the quotients of the $g_{\mu\nu}$ are determined by light signals. The following consideration will make this clear. Let a metric $g_{\mu\nu}$ be determined by means of rigid rods and clocks; in order to discover whether in this world light satisfies axioms I to V, it must be investigated whether the world lines $ds^2 = 0$ satisfy these axioms. If this is the case, then every metric $g'_{\mu\nu}$ which results from $g_{\mu\nu}$ through multiplication by a scalar field $\lambda(x_1 \ldots x_4)$ will satisfy these axioms, since in it the lines $ds^2 = 0$

satisfy the same equation. Accordingly, the most general metric satisfying axioms I to V will have the form

$$\begin{vmatrix} \lambda & 0 & 0 & 0 \\ 0 & \lambda & 0 & 0 \\ 0 & 0 & \lambda & 0 \\ 0 & 0 & 0 & -\lambda \end{vmatrix} \tag{97}$$

If a metric of this form resulted from a measurement with rigid rods and clocks, it would be different from the light geometry. The light geometry would be Euclidean, because the introduction of the light-geometrical metric determination (definitions 7 and 10) would be equivalent to the division of ds^2 by λ.

Such a world is mathematically possible, but it is shown to be physically impossible by the theory of relativity on the basis of another principle. In order to understand this fact, we must distinguish two cases. Either the metric (97) satisfies Einstein's gravitation equation of the vacuum $R_{\mu\nu} = 0$ or it does not. Let us consider the first case. According to a theorem proved by Schouten and Struik,[44] the metric is Euclidean, that is, it can be transformed by a transformation of coordinates into the normal schema (77). This transformation belongs to the spherical transformations mentioned in § 16 and has a singularity. We have now obtained two descriptions of the world: in one of them the light geometry and the matter geometry are identical and Euclidean, but a singularity exists; in the other one the light geometry is Euclidean, the matter geometry is non-Euclidean, and no singularity exists. It is not along the lines of the special theory of relativity to admit these possibilities. On the contrary, one would assume that exactly where light geometry and matter geometry are identical and Euclidean the singularity is missing, and not the other way around. In this case the state of affairs asserted by us in § 16 exists and

[44] J. A. Schouten and D. J. Struik, *American Journal of Mathematics*, Vol. XLIII, no. 4 (October 1921). A metric of the form (97) can be mapped conformally upon a Euclidean one. The corresponding theorem reads: If for such a conformal Euclidean metric $R_{\mu\nu} = 0$, then it is Euclidean, that is, $R_{iklm} = 0$.

the validity of axioms VII to X is connected with the singularity-free validity of axioms I to V. Let us now consider the second case, $R_{\mu\nu} \neq 0$. In such a world a difference between matter geometry and light geometry exists that cannot be transformed away. But this world is incompatible with Einstein's field equations of the vacuum which require $R_{\mu\nu} = 0$ and, therefore, it is physically impossible although mathematically possible. There is no such world.[45] Only here can one recognize clearly why the motion of light determines the absolute values in the case of a Euclidean geometry even though it determines only the quotients of the $g_{\mu\nu}$: the case of an added scalar factor would result in a gravitational field that would be physically impossible.

Quite generally it can be maintained that the validity of axioms VII to X can be derived from the validity of axioms I to V, that is, that the inertial system is determined by the properties of light alone.

But this statement is correct only if regions of finite extension are concerned. For local regions, fields of the form (97) are not exluded, because if this form holds only infinitesimally, then the derivatives $\frac{\partial g_{\mu\nu}}{\partial x_\sigma}$ and $\frac{\partial^2 g_{\mu\nu}}{\partial x_\sigma \partial x_\rho}$ are not tied to the forms $\frac{\partial \lambda}{\partial x_\sigma}$ and $\frac{\partial^2 \lambda}{\partial x_\sigma \partial x_\rho}$, and therefore the requirement $R_{\mu\nu} = 0$ can still be satisfied. The application of the light geometry in the infinitesimal would furnish only local fields of the form (97) and not contain any means of distinguishing the coordinate system obtained in this way from those in which the $g_{\mu\nu}$ have the normal $\overline{g_{\mu\nu}}$. Light-geometrically, only the quotients $q_{\mu\nu} = \frac{g_{\mu\nu}}{-g_{44}}$ can be determined.

---

[45] For matter-filled spaces, that is, for $R_{\mu\nu} \neq 0$, a metric (97) is possible. But here light does not satisfy the equation $ds^2 = 0$ and therefore the light geometry cannot be carried through. I have not yet been able to decide whether the field equations extended by the cosmological term would exclude solution (97); it is probable, however.

The same is true for the determination of the local inertial systems; they cannot be obtained light-geometrically either. We have shown that these systems are characterized by the minimal requirement that the deviation from the axioms should be a minimum. For the light axioms this requirement is satisfied if $\frac{\partial g_{\mu\nu}}{\partial x_\sigma} = 0$, but this condition is not necessary; it suffices for the validity of the light axioms that $\frac{\partial q_{\mu\nu}}{\partial x_\sigma} = 0$. This yields

$$\frac{\partial g_{\mu\nu}}{\partial x_\sigma} = \frac{g_{\mu\nu}}{g_{44}} \cdot \frac{\partial g_{44}}{\partial x_\sigma}. \tag{98}$$

The local systems corresponding to this requirement form a more general class containing the local inertial systems.

If the light geometry furnishes only the quotients of the $g_{\mu\nu}$, it is not without value for the construction of the metrical determination. The $q_{\mu\nu}$ obtained in this way in a coordinate system have a definite meaning for this system. The same $q_{\mu\nu}$ would be obtained if the $g_{\mu\nu}$ were first produced by means of rigid rods and clocks and then divided by $-g_{44}$. We shall show later how this idea can be used for the construction of the metrical determination; it has special practical significance in the static gravitational field. The $q_{\mu\nu}$ have definite values for every coordinate system; their law of transformation is

$$q'_{\mu\nu} = -\frac{\dfrac{\partial x_i}{\partial x'_\mu} \cdot \dfrac{\partial x_k}{\partial x'_\nu}\, q_{ik}}{\dfrac{\partial x_l}{\partial x'_4} \cdot \dfrac{\partial x_m}{\partial x'_4}\, q_{lm}}. \tag{99}$$

This transformation is not tensorial; the $q_{\mu\nu}$ do not form a tensor.

For the determination of the absolute values of the $g_{\mu\nu}$ in the gravitational field, however, we need material things,[46]

[46] Once again the path of a force-free mass point may take the place of a material thing, as H. Weyl has shown. The position of a geodesic line in the coordinate system is changed with a change of calibration; therefore, the calibration is determined as soon as this position is given (cf. note on p. 82).

either natural clocks or rigid rods. Herein lies the difference between the general and the special theory. We must therefore discuss this definition of the metric in the following section.

§ 37. *The Characteristic Length of a Unit-Measuring Rod*

We shall now turn to the standardization of the ds^2 with material objects. It is customary to say that $ds^2 = 1$ everywhere for a unit clock and a rigid rod. This is correct for the clock, because the beginning and end of the period are two events and for such events a ds^2 exists. It is different for a measuring rod. It determines only space points and in order to obtain the ds^2, one must first produce two events. These two events must happen simultaneously at the ends of the rod, since the problem concerns a measurement of length; dx_4 is to be $= 0$. Since simultaneity is arbitrary, the ds^2 of the rod is not defined before it is indicated how simultaneity will be defined. Thus we arrive at the following definition.

Definition 26. The "characteristic length" of a measuring rod is obtained if two simultaneous events happen at its ends whose simultaneity is defined in such a way that the velocity of light along the rod in both directions is equal.

This requirement can be tested by means of two light signals which meet in the middle of the rod. If a subdivision of the rod is to be avoided, the following method can be used. The two signals meet at a certain point P, continue, are reflected at the ends of the rod, and meet a second time. If they meet once more at P, the events were simultaneous. It is also possible to establish clocks at the ends of the rod which are synchronized according to definition 8.

The significance of definition 26 is as follows. Let a rigid rod be given in any coordinate system. Definition 26 has the effect that with this choice of local simultaneity the $g_{\alpha 4} = 0$ ($\alpha = 1, 2, 3$), because for $ds^2 = 0$

$$dx_4 = \frac{-d\gamma}{g_{44}} \pm \sqrt{\frac{d\gamma^2}{g_{44}^{\ 2}} - \frac{d\sigma^2}{g_{44}}} \qquad \begin{aligned} d\sigma^2 &= g_{\alpha\beta}\, dx_\alpha\, dx_\beta \\ d\gamma &= g_{\alpha 4}\, dx_\alpha \\ \alpha, \beta &= 1, 2, 3. \end{aligned} \qquad (100)$$

If the sign of the dx_α is reversed, that is, if a light signal is sent in the opposite direction at the same space point at the same time, then $d\gamma$ changes its sign, whereas the terms under the radical retain their values as quadratic forms. Therefore, dx_4 will have a different value. Only if $d\gamma = 0$, that is, if $g_{\alpha 4} = 0$, since the dx_α are arbitrary, will dx_4 be the same for both directions. A choice of simultaneity for which dx_4 becomes equal in both directions of the spatial line element is equivalent to $g_{\alpha 4} = 0$.

This simultaneity is characterized by the fact that with conservation both of rest for a rod and of simultaneity a transition can be made to a local orthogonal system. It can further be shown that this simultaneity is a *necessary* condition for this transition. Definition 26 is adjusted in such a way that it corresponds to the rule of the special theory of relativity for that infinitesimal system in which the rod is at rest.

If by means of a transformation of the form (82) with (83) and (85) a transition is made to a local orthogonal system, the following simplified equations of restriction for the first derivatives are obtained:[47]

$$\left.\begin{aligned} g'_{\alpha\beta} &= \delta_{\alpha\beta} = \frac{\partial x_\gamma}{\partial x'_\alpha}\frac{\partial x_\delta}{\partial x'_\beta}\, g_{\gamma\delta} \\ g'_{44} &= -1 = \left(\frac{\partial x_4}{\partial x'_4}\right)^2 g_{44} \\ g'_{\alpha 4} &= 0 = \frac{\partial x_\gamma}{\partial x'_\alpha}\frac{\partial x_4}{\partial x'_4}\, g_{\gamma 4} \end{aligned}\right\} \alpha, \beta, \gamma, \delta = 1, 2, 3. \qquad (101)$$

The last three equations $g'_{\alpha 4} = 0$ in which $\dfrac{\partial x_4}{\partial x'_4}$ may be canceled are three homogeneous linear equations for the three magnitudes $g_{\gamma 4}$; they can be satisfied only if either $g_{\gamma 4} = 0$ or if the determinant of their coefficients vanishes. But this determinant is identical with the subdeterminant D_4^4 of transformation

[47] Cf. note on p. 83 for $\delta_{\alpha\beta}$.

(82) and since the whole determinant D reduces to $\frac{\partial x_4}{\partial x_4'} \cdot D_4^4$ because of (83) or (85), respectively, D would also vanish in this case. Since the transformation is to be infinitesimally unique, this case must be excluded and, therefore, $g_{\gamma 4} = 0$ is the necessary condition for the possibility of going to the local orthogonal system upon a conservation of the simultaneity and of rest for a rod.

One can always go from the local orthogonal system to the local inertial system while simultaneity and rest for a rod are preserved, because the requirement $\frac{\partial g_{\mu\nu}}{\partial x_\sigma} = 0$ can always be satisfied by the second derivatives alone. One can therefore say that the characteristic length determined with definition 26 is identical with the characteristic length that the same rigid rod would have in the local orthogonal system and in the inertial system; in this fact lies the justification of definition 26.

§ 38. *The Construction of the Metric in the General Case*

One can now imagine the construction of the metric $g_{\mu\nu}$. If any coordinate system K is given for the world, then point events of the distance $ds^2 = +1$, $ds^2 = 0$, $ds^2 = -1$ can be produced by means of rigid rods and clocks permanently at rest in K. The world will thus be interspersed with unit mesh points. The functions $g_{\mu\nu}$ are to be determined in such a way that with the chosen coordinates $ds^2 = \pm 1$ or 0, respectively, for all mesh points.

If merely the quotients $q_{\mu\nu}$ are to be constructed, light signals alone suffice. Only mesh points $ds^2 = 0$ are constructed and the $q_{\mu\nu}$ are determined in such a way that with the chosen coordinates $ds^2 = 0$ for these mesh points.

In addition to this indirect procedure, there exists a direct procedure for the construction of the $g_{\mu\nu}$; in this case, separate local systems K' must be used in addition to the general system

K. Local coordinate systems with an orthogonal metrical determination are constructed in the environment of a point *Q* according to the rules of axioms I to X and definitions 1 to 21; the transformation connecting *K′* and *K* is determined at *Q* and the $g_{\mu\nu}$ in *K* are calculated thereby at the point *P*. This must be carried through separately for every point *Q*. For this purpose the light geometry in *K′* suffices. Only in order to make the unit of length (or the unit of time) of the various systems *K′* equal are transportable rigid rods (or clocks) needed. The significance of material things becomes clear: they bring about a comparison of the units at different points. This comparison cannot be achieved by means of light signals; for all other purposes light is sufficient.

In this procedure it is preferable to use local inertial systems for *K′*, because for the same extension of *K′* the given method of approximation will become more exact. But this is the only advantage of the local inertial systems; they are not necessary for the direct construction.

We must now demonstrate the uniqueness of this procedure. Although the constructed $g_{\mu\nu}$ will be different for every coordinate system, it must be irrelevant whether the $g_{\mu\nu}$ are directly determined for a local system *K′* or whether they are first determined for another local system *K″* and then transformed to *K′*.

We recall the investigations of the first part which have shown that the metrical determination is always the same, independently of the question which of the uniformly moving inertial systems is the starting point. The conditions for this result are already contained in axioms I to X and are specifically formulated by axioms IX and X.

Axiom XI,2 states in addition that, given the starting point *P*, the metrical determination is always the same, no matter what acceleration the clocks and measuring rods have as long as they are at rest infinitesimally. If one goes now to a starting point with a different velocity in the same world region, then according to axiom XI,2 the same result holds with respect to this point *P′*. At *P* and *P′* clocks and rods can be used so that they are in each case infinitesimally at rest for P or P', respectively, but have accelerations so that they are moving uniformly relative to one another. According to our discussion above, they lead to the same metrical determination, and therefore any systems of different acceleration connected with P and P' must furnish the same metrical determination. Therefore, definition 24 is uniquely satisfied.

Chapter Four

Integral Properties

§ 39. *The Problem*

After the construction of the metric in the general case another problem arises which does not exist in the special theory, namely, the problem of integral properties. Until now we have characterized space and time only by means of a differential principle. Now the question arises: Are some of the axioms I to X valid for finite extended regions as a consequence of the validity of the totality of these axioms in infinitesimal regions? The answer to this question will clarify what space and time mean in the general case.

At first, we shall not treat this question in its most general form, but begin with special cases. Schematically this must be understood as follows. If it should turn out to be an empirical fact that the $g_{\mu\nu}$ satisfy certain simple conditions (for instance, those of the static field), when the metric is constructed according to § 38, what axioms will be valid in a finite region? Conversely, the validity of these axioms in a finite region is a criterion of this special case. In this way it is possible to make the transition from the special Euclidean space-time world to the general one in steps. For this purpose, our construction of the axioms will turn out to be advantageous, because we shall find that every elimination of a light axiom in the finite region will constitute one step. At the end, the general signal axioms I and II will exhibit the properties of the general space-time world which can be characterized only topologically, no longer metrically.

The difference between "finite region" and "the world as a whole" will not be essential at the outset, since we may assume that the restrictions for the $g_{\mu\nu}$ hold for the whole world. Only at the last step where these restrictions do not apply will this difference be significant again.

§ 40. *Static Fields*

We shall begin with the static field. It is defined by these requirements: $g_{\mu\nu}$ independent of x_4, $g_{\alpha 4} = 0$, $d\sigma^2 = g_{\alpha\beta}\, dx_\alpha\, dx_\beta$ ($\alpha, \beta = 1, 2, 3$) positive definite, $g_{44} < 0$. The last two requirements do not indicate a peculiarity of the static field as such but merely that the static field shall belong to real systems; we shall therefore postpone the discussion of the significance of these requirements to § § 46 and 47. There we shall show that these requirements imply the validity of axioms I and II.

Axiom III is also valid. This axiom formulates Fermat's principle of the earliest arrival which holds for static fields.[48] Furthermore, axiom IV,1 is valid, because the $g_{\mu\nu}$ are independent of the time, that is, because the time $\int dx_4$ always furnishes the same value for a light signal along a certain path. In addition, axiom IV,2 is valid. This is a consequence of the fact that the time dx_4 is equal for both directions of the path over a spatial line element $d\sigma$. From $ds^2 = 0$ it follows that

$$dx_4 = \frac{d\sigma}{\sqrt{-g_{44}}}, \quad d\sigma^2 = g_{\alpha\beta}\, dx_\alpha\, dx_\beta\,, \quad \alpha, \beta = 1, 2, 3\,. \quad (102)$$

If the sign of the spatial differentials dx_α and dx_β is reversed, that is, if the light signal is sent in the opposite direction, then $d\sigma^2$ as a homogeneous quadratic form changes neither value nor sign; therefore, dx_4 remains unchanged.

---

[48] Cf., for instance, Pauli, *Relativitätstheorie* (Enzyklopädie), p. 716.

Axiom V, however, no longer holds. The non-Euclidean character of the light geometry shows that a gravitational field exists; this is the difference between the static field and the inertial system. But the light geometry can still be carried through because of the validity of axioms I to IV. Definition 10 of the equality of distances leads now to a metrical determination different from Einstein's, because in combination with definition 8 of simultaneity it furnishes the result that the velocity of light becomes constant $= 1$ in the whole system. Since, on the other hand, the velocity of light c^* in the static case becomes $= \frac{d\sigma}{dx_4} = \sqrt{-g_{44}}$, g_{44} is set $= -1$, that is, not Einstein's $g_{\mu\nu}$ but the $q_{\mu\nu} = \frac{g_{\mu\nu}}{-g_{44}}$ are obtained. This result explains why definition 9 of a straight line as the path of a light ray is justified. In the static field of the Einsteinian metrical determination, light, although moving on four-dimensional geodesic zero lines, is not moving on three-dimensional geodesics.[49] According to Fermat's principle, $\int \frac{d\sigma}{\sqrt{-g_{44}}}$, not $\int d\sigma$, becomes an extremum. In the metrical determination of the $q_{\mu\nu}$, however, these two integrals become identical. This is the reason that in the light geometry of the static field light rays are the straightest lines of three-dimensional space.

Since the spatial geometry does not become Euclidean, it is impossible to define the static field uniquely in terms of the light geometry. Already in § 13 we encountered two systems having a common point, both of which satisfy axioms I to IV and only one of which is rigid. Both of them have static $q_{\mu\nu}$, but only the system K also satisfying axiom V has static $g_{\mu\nu}$, too, whereas the system K', being in uniform radial expansion, possesses nonstatic $g'_{\mu\nu}$.

The metric of K' is obtained if transformation (33) is applied to the Euclidean metric for polar coordinates. If the polar

[49] Cf. W. Pauli, *op. cit.*, and *PST*, p. 72 n.

coordinates ρ, ϑ, ψ of the expanded system K' are designated by x'_1, x'_2, x'_3 and its time τ by x'_4, then

$$q'_{\mu\nu} = \begin{matrix} 1 & 0 & 0 & 0 \\ 0 & \frac{1}{4}(e^{x'_1} - e^{-x'_1})^2 & 0 & 0 \\ 0 & 0 & \dfrac{\cos^2 x'_2}{4}(e^{x'_1} - e^{-x'})^2 & 0 \\ 0 & 0 & 0 & -1 \end{matrix}$$

and for the $g'_{\mu\nu}$

$$g'_{\mu\nu} = \begin{matrix} e^{2x_4} & 0 & 0 & 0 \\ 0 & \frac{1}{4}(e^{x'_4+x'_1} - e^{x'_4-x'_1})^2 & 0 & 0 \\ 0 & 0 & \dfrac{\cos^2 x'_2}{4}(e^{x'_4+x'_1} - e^{x'_1-x'_1})^2 & 0 \\ 0 & 0 & 0 & -e^{2x'_4}. \end{matrix}$$

This system retains the peculiarity $R'_{iklm} = 0$; it is Euclidean, since there is a transformation to the orthogonal schema of the system K. There are also fields for which $R_{iklm} \neq 0$ and whose $q_{\mu\nu}$ nevertheless satisfy axioms I to IV. Here we cannot reason by analogy to § 36; since the $g_{\mu\nu}$ are not bound to the form (97) by the validity of axiom V for the $q_{\mu\nu}$, these fields can still satisfy the condition $R_{\mu\nu} = 0$. The light geometry leaves the metric undetermined even in the static case up to a scalar field.

The validity of axioms I to IV determines a wider class than Einstein's concept of the static system; they might be called light-static systems. If among the systems of this class the static system is to be defined specifically, a rigid rod or a natural clock must be added. If a clock is used, then that system is static in which axioms I to IV are valid and in which the clock runs uniformly in the sense of the natural clock. The rigid rod could be used for the construction of a light clock and then handled in the same manner.

Time has very simple properties in a static system. The synchronism is symmetrical and transitive. Imagine that a central clock has been set up from which all clocks are adjusted according to the definition $t_2 = \frac{t_1 + t_3}{2}$; then all of them are synchronous relative to one another. The periods are uniform, that is, determined by this property up to a linear function. The only difference from inertial systems consists in the fact that the time unit cannot be transported by means of natural clocks; it is dependent on the location.

Since the light geometry can be carried through in the static field, there exists a direct procedure for the construction of the metric which needs neither mesh points nor local systems according to § 38. Light-geometrically, one first obtains the $q_{\mu\nu}$; the time is adjusted according to definition 8. If a natural clock is added, g_{44} is obtained by a transport of this clock to the different real points of the system. The relation of the time unit, transmitted by signals, to the unit of the natural clock is at every point directly $= \sqrt{-g_{44}}$ if originally $g_{44} = -1$. Through $g_{\mu\nu} = -q_{\mu\nu} \cdot g_{44}$ the $g_{\mu\nu}$ are obtained.

The reason for this simple determination of the g_{44} is the fact that the time difference (measured in the coordinate time of the system) between two light signals sent one after the other from a certain point is equal at every other point of a static system which they encounter on their path. For the light ray

$$dx_4 = \sqrt{\frac{d\sigma^2}{-g_{44}}}$$

is independent of the time if the $g_{\mu\nu}$ are independent of the time. (It is sufficient if the $q_{\mu\nu}$ are independent of the time.) If a light ray travels from P_1 at the time $x_4^{(1)}$ to a point P_2 which has the distance $\int_l d\sigma$ from P_1 (measured on the path l of the light ray in three-dimensional space), it will arrive at P_2 at the time $x_4^{(2)} = x_4^{(1)} + \int_l dx_4 = x_4^{(1)} + \int_l \sqrt{q_{\alpha\beta}dx_\alpha dx_\beta}$. A signal which is sent from P_1 at a later time $X_4^{(1)}$ will arrive at the time $X_4^{(2)} = X_4^{(1)} + \int_l \sqrt{q_{\alpha\beta}dx_\alpha dx_\beta}$ at P_2. Since the $q_{\alpha\beta}$ are independent of the time x_4, $X_4^{(2)} - X_4^{(1)} = x_4^{(2)} - x_4^{(1)}$. In a static system the time coordinate is chosen from the outset in such a manner that the time difference $(X_4 - x_4)$ of two events at one point will remain constant if it is projected upon another point by means of light rays. If a natural unit clock U_1 is set up at P_1, then the interval between the beginning and the end of its period $= 1$, and therefore $ds = 1$ (definition 24 and axiom XI, 2). To this interval belongs an interval of the coordinate time $dx_4^{(1)}$ which results in $dx_4^{(1)} = \dfrac{1}{\sqrt{g_{44}^{(1)}}}$ (since $dx_1 = dx_2 = dx_3 = 0$ for the clock at rest). When it is projected upon P_2, an interval of coordinate time $dx_4^{(2)}$ results which is $= dx_4^{(1)}$ according to the above consideration. However, a natural clock U_2 set up at P_2 furnishes there an interval whose coordinate time is $\overline{dx_4^{(2)}} = \dfrac{1}{\sqrt{g_{44}^{(2)}}}$. Therefore, $dx_4^{(2)} \neq \overline{dx_4^{(2)}}$, that is, $\dfrac{dx_4^{(2)}}{\overline{dx_4^{(2)}}} = \dfrac{\sqrt{g_{44}^{(2)}}}{g_{44}^{(1)}}$. If the light signals which arrive at P_2 are timed at P_2 according to the periods of the clock U_1, and if they are compared there with U_2, then the ratio of the times furnishes directly the square root of the ratio of the g_{44} for the two points. (Upon projection by light rays, the interval of coordinate time dx_4, not the ds, is preserved.)

The special properties of the time of a static field can furthermore be characterized in the following way. In a static field a transformation can be made at every point to a local inertial system, so that rest for a rod, simultaneity, uniformity, and temporal sequence are preserved. It was shown in § 37 that a system in which the $g_{\alpha 4}$ vanish changes into an inertial system while rest for a rod and simultaneity are preserved. The vanishing of the $\dfrac{\partial g_{\mu\nu}}{\partial x_4}$ is the reason that in the static case uniformity and temporal sequence are also preserved. If in transformation (81) the indices $\mu = \nu = \sigma = 4$ are set for

$\frac{\partial g'_{ik}}{\partial x'_l}$, this equation is reduced, because of (85), $\frac{\partial g_{\mu\nu}}{\partial x_4} = 0$, and $g_{\alpha 4} = 0$ (x'_4 coordinate of the local inertial system), to

$$0 = 2\frac{\partial^2 x_4}{\partial x_4'^2} \cdot \frac{\partial x_4}{\partial x'_4} g_{44}.$$

Here $g_{44} \neq \frac{\partial x_4}{\partial x'_4}$ must not vanish, since otherwise the whole determinant of the transformation would vanish because of (85). Therefore,

$$\frac{\partial^2 x_4}{\partial x_4'^2} = 0.$$

This expression states the conservation of uniformity. Since, in addition, rest for a rod and simultaneity are preserved, the conservation of uniformity is symmetrical.

In this sense it can be said that a static field has the same time as the special theory of relativity.

We shall now consider the matter axioms. In a gravitational field rigidity is defined only infinitesimally. But the rigidity of finite things can be defined as follows.

Definition 27. A spatial coordinate system is called rigid if any two infinitesimally neighboring points can be connected permanently by a rigid rod.

In this sense the static system is rigid and axioms VII and VI, 1 are valid in finite regions. All other matter axioms are valid only infinitesimally. We should like to demonstrate this result particularly for axiom VIII. Since in a static field simultaneity is the same as in the inertial field, the characteristic length of a rigid rod will be $= d\sigma$.[50] Therefore, two rods

[50] Here lies the justification of Einstein's conclusion concerning "the behavior of measuring rods and clocks in static gravitational fields" (*Grundlage der allgemeinen Relativitätstheorie* [1916], p. 60). This inference can be applied only to static fields; cf. § 44. [A translation of this work is contained in *The Principle of Relativity: Original Papers* by A. Einstein

$P_1 P_2$ and $P_1 P_3$ have the same length if $\int_{P_1}^{P_2} d\sigma = \int_{P_1}^{P_3} d\sigma$. But then the time for a light signal will be $\overline{P_1 P_2} \neq \overline{P_1 P_3}$, since $\int_{P_1}^{P_2} \frac{d\sigma}{\sqrt{-g_{44}}} \neq \int_{P_1}^{P_3} \frac{d\sigma}{\sqrt{-g_{44}}}$ because g_{44} is different along these two paths. Since $\overline{P_1 P_2 P_1} = 2\, \overline{P_1 P_2}$, $\overline{P_1 P_2 P_1} \neq \overline{P_1 P_3 P_1}$. In a gravitational field the Michelson experiment would therefore have results different from those in the inertial system if along the arms of the apparatus a noticeable change of the g_{44} occurred, that is, if the arms were large relative to the local $d\sigma$. If the arms are of the magnitude $d\sigma$, the result is the same as in the inertial system, because the deviation arises only from the change of the g_{44} with location. Axiom VIII holds only in infinitesimal regions for neighboring points. Even in a static field, the light clock must have infinitesimal dimensions, because otherwise its unit would be dependent upon its orientation.

§ 41. *A Time Definition Whose Unit Is Transportable*

In a static system S the time unit is not transportable and can be transmitted only by signals. The question arises whether a time definition x_4' can be carried through that would avoid this disadvantage. This is possible. One defines as time unit for every point the unit of the clock transported there. The synchronism can still be adjusted, for instance, in such a way that (defined from S) it is to be identical at a certain moment with the synchronism of S (here $x_4' = 0$); then it will be determined for all later times by the periods of the established clocks, because equal figures on the clocks mean simultaneous events. The result would be a simultaneity that contradicts our intuition and is inadequate; simultaneous events according to x_4',

and H. Minkowski, translated by M. N. Saha and S. N. Bose, with an historical introduction by P. C. Mahanalobis. Calcutta: University of Calcutta, XXIII.—*Ed.*]

measured according to x_4, will at later times no longer be simultaneous. The intervals between them would become larger and larger. A light ray going from P_1 to P_2 will obtain there time coordinates $x_4'^{(1)}$ and $x_4'^{(2)}$ so that the difference $x_4'^{(2)} - x_4'^{(1)}$ will become larger and larger at later times. The velocity of light is therefore variable with respect to time. The system S' constructed by means of the same spatial coordinates $x_1x_2x_3$ as S and the time x_4' is therefore no longer static. But it is an admissible coordinate system, and even its simultaneity is logically and mathematically admissible. It is obtained by means of the coordinate transformation

$$x_1' = x_1 \quad x_2' = x_2 \quad x_3' = x_3 \quad x_4' = \sqrt{-g_{44}} \cdot x_4 .$$

The time coordinate is multiplied everywhere by a factor dependent on the location.[51] $g_{44}' = -1$, but the spatial $g_{\mu\nu}'$ are dependent on x_4', because the $\dfrac{\partial x_4}{\partial x_\alpha'}$ $(\alpha = 1, 2, 3)$ contain the time x_4'.

§ 42. *Red Shift, Deflection of Light, Perihelion of Mercury*

The proof of the preservation of the interval of coordinate time under a projection by light rays given in § 40 is also the derivation of the red shift which lies at the foundation of Einstein's presentation of this problem. The red shift depends on the following assumptions.

1. The atom is a clock (cf. § 35).
2. Light satisfies the equation $ds^2 = 0$.[52]
3. The world is static (or stationary) with sufficient approximation.

---

[51] g_{44} is dependent only on $x_1\ x_2\ x_3$, since it belongs to a static system.

[52] M. v. Laue's calculations (*Relativitätsprinzip II*, p. 190) are not necessary for a justification of the red shift if it is assumed that light moves on zero lines. But these calculations are valuable, because they have shown that this assumption, derivable from the differential principle of the theory, is at the same time derivable from the extrapolated Maxwellian equations; thus they have demonstrated the internal consistency of the theory. It would be sufficient for the red shift if light moved not on zero lines but in a stationary manner, that is, if it always described the same spatial path and only had a spatially, not temporally, variable velocity.

4. The $g_{\mu\nu}$ must transform with progressive approximation into the Newtonian gravitation potential according to Einstein's method.

The last of these assumptions is of special significance. It is to be noted that the values of the $g_{\mu\nu}$ for the astronomic world are not obtained directly through the procedure described in §§ 38 or 40, but inferred indirectly from it, and that they are to achieve an approximate transition to Newton's theory. The method used by Einstein is very plausible, but there is no unique rule; essentially, it contains as assumptions the field equations of gravitation and the principle of equivalence.

Yet the metrical behavior of clocks relative to light is of secondary significance. Even if the clock does not adjust to the light geometry, but does adjust in the sense of absolute time according to axiom *B* of § 23, the red shift follows, only quantitatively changed by magnitudes of second order. This result is due to the fact that the red shift can also be calculated as a Doppler effect according to the principle of equivalence (two clocks at rest in an accelerated system have a Doppler effect relative to each other, because their velocity, relative to an inertial system, increases constantly). Axiom *B* and axiom X differ relative to the Doppler effect only by magnitudes of second order. Only axiom *A* is excluded by the red shift, because the transport of clocks through gravitational fields would change their periods.[53]

Essentially, the red shift depends not on an assumption about the metrical behavior of the clocks relative to light, but on an assumption about their metrical behavior relative to gravitation. It is assumed that the standardization of the $ds^2 = 1$ by means of the clocks leads just to those $g_{\mu\nu}$ standing in a certain simple relation to the phenomena of gravitation. (This is the equivalence principle.)

A similar phenomenon is the bending of light rays; it shows that Einstein's simple relation exists between the quotients of the $g_{\mu\nu}$ determined by the motion of light, that is, between

[53] At least in fields which cannot be "transformed away."

the $q_{\mu\nu}$ and gravitation. The bending of the light rays depends on assumptions 2, 3, and 4. However, since light determines only the quotients $q_{\mu\nu}$, the deflection of light proves nothing about the absolute values of the $g_{\mu\nu}$; it can tell nothing about whether the standardization of the $ds^2 = 1$ by means of clocks and measuring rods is compatible with the theory of gravitation.

The third empirical test of the general theory is the motion of the perihelion of Mercury. It supports merely the correctness of the theory of gravitation obtained from a tensorial potential, but proves nothing concerning the relation of these $g_{\mu\nu}$ to clocks, rods, and light. It contains only assumptions 3 and 4.

Summarizing, we can say that only the deflection of light and the red shift can furnish evidence for Einstein's relation between gravitation and the real metric defined according to axioms I to XI; only the red shift shows that the standardization of the $ds^2 = 1$ by clocks is compatible with this relation.

§ 43. *Stationary Fields*

The properties of these fields are given by these requirements: $g_{\mu\nu}$ independent of x_4, but $g_{\alpha 4} \neq 0$ ($\alpha = 1, 2, 3$), $d\sigma^2$ positive definite, $g_{44} < 0$.

In these fields axioms I, II, and IV,1 are valid in finite regions, whereas axiom IV,2 no longer holds, since for $ds^2 = 0$

$$dx_4 = -\frac{d\gamma}{g_{44}} + \sqrt{\frac{d\gamma^2}{g_{44}^{\;2}} - \frac{d\sigma^2}{g_{44}}} \qquad \begin{aligned} d\gamma &= g_{\alpha 4}\, dx_\alpha, \quad \alpha, \beta = 1, 2, 3 \\ d\sigma^2 &= g_{\alpha\beta}\, dx_\alpha\, dx_\beta . \end{aligned} \tag{103}$$

The sign of the radical must be positive. $d\gamma$ changes its sign if the dx_α change their signs, that is, if the spatial line element is traveled in the opposite direction. The terms under the radical do not change. Therefore, dx_4 has a different numerical value and axiom IV,2 cannot hold.

Only in special instances will the various dx_4 combine along each path of the triangle in such a way that the round-trip times will nevertheless be equal. In this case the stationary system can be transformed into a static one by a pure time transformation, because only definition 8 of simultaneity has to be carried through. It is an "apparently stationary" system. Such systems can be characterized by a condition for the $g_{\mu\nu}$. If the stationary system K is to be transformed into a static system K' by a transformation

$$\begin{aligned} x_\alpha &= x'_\alpha \\ x_4 &= f(x'_1 \cdots x'_4), \end{aligned} \tag{104}$$

then

$$g'_{\alpha 4} = \frac{\partial x_i}{\partial x'_\alpha}\frac{\partial x_k}{\partial x'_4}\, g_{ik} = 0 \tag{105}$$

must hold.

Because of the special character of transformation (104), we must set $k = 4$, since the other terms vanish. Furthermore, i furnishes expressions different from 0 only for the values $i = \alpha$ and $i = 4$; $\dfrac{\partial x_\alpha}{\partial x'_\alpha}$ will be $= 1$. If one divides by $\dfrac{\partial x_4}{\partial x'_4}$ which does not vanish, then (105) yields

$$\frac{\partial x_4}{\partial x'_\alpha} = -\frac{g_{\alpha 4}}{g_{44}}. \tag{106}$$

From this expression we obtain a condition for the g_{i4} by differentiation. Therefore, since $\dfrac{\partial}{\partial x'_\alpha} = \dfrac{\partial}{\partial x_\alpha}$,

$$\frac{\partial x_4}{\partial x'_\alpha \partial x'_\beta} = -\frac{\partial}{\partial x_\beta}\left(\frac{g_{\alpha 4}}{g_{44}}\right) = -\frac{\partial}{\partial x_\alpha}\left(\frac{g_{\beta 4}}{g_{44}}\right).$$

Calculation furnishes a condition

$$g_{44}\left(\frac{\partial g_{\beta 4}}{\partial x_\alpha} - \frac{\partial g_{\alpha 4}}{\partial x_\beta}\right) - g_{\beta 4}\frac{\partial g_{44}}{\partial x_\alpha} + g_{\alpha 4}\frac{\partial g_{44}}{\partial x_\beta} = 0, \tag{107}$$

which the $g_{\mu\nu}$ of the stationary field must satisfy so that the field can be transformed into a static one by a mere time transformation; the calculation furnishes also a condition for the validity of the round-trip axiom IV,2 in a stationary field.

Condition (107) is necessary but not sufficient because of the additional requirement that in K' also $\frac{\partial g'_{\mu\nu}}{\partial x'_4} = 0$. This leads to ten equations for the eight magnitudes $\frac{\partial x_4}{\partial x'_\mu}$ and $\frac{\partial^2 x_4}{\partial x'_\mu \partial x'_4}$ to which is added the equation that results from (106) by differentiation with respect to x_4. These eleven equations can be solved for the $g_{\mu\nu}$ only under certain conditions. If (107) alone is satisfied, this means that the $g_{\alpha 4}$ can be made to vanish by a pure time transformation, but that the resulting system does not possess $g_{\mu\nu}$ independent of time. In this respect (107) is the most important condition.

Since the round-trip axiom is no longer valid, the synchronism in the stationary field is merely symmetrical, no longer transitive, and "simultaneity" does not mean the same as in the inertial system. Upon transformation on the local inertial system only uniformity and the stationary state of a rod, not simultaneity, are preserved. This is the reason the $d\sigma$ of the system is not identical with the characteristic length of a rigid rod; it is impossible to infer the geometry of rigid rods directly from the geometry of the $d\sigma$. Since these relations are not quite clear, we shall investigate them more closely by means of the rotating circular disk.

§ 44. *The Rotating Circular Disk*†

A very simple definition of time exists for the rotating circular disk. A clock is set up at the center and from there all clocks are synchronized by means of direct light signals according to $t_2 = \frac{t_1 + t_3}{2}$. Because of the validity of axiom VI, 1 the synchronism between the central clock and all other clocks becomes symmetrical but, as we shall show, it is no longer transitive. Since, however, as we shall also show, no transitive synchronism exists at all for the circular disk,

† See editor's reference notes, p. 204.

the above definition of time is still the most natural one.[54]

This definition of time has the effect that every clock on the circular disk K always shows the same time as that particular clock of the stationary plane $\bar{\bar{K}}$ over which it happens to be located at that moment. The signal of adjustment is identical for these two clocks for both directions of each path. The time transformation from $\bar{\bar{K}}$ to K is therefore $\bar{\bar{x}}_4 = x_4$.

Through the requirements that

1) The definition of time of the circular disk is prescribed, that
2) every point of the circular disk rotates with the same angular velocity, and that
3) the metrical determination of the circular disk is everywhere infinitesimally the same as Einstein's,

the coordinate transformation and the metric of the circular disk are completely determined. Requirement 2 determines the stationary points of the system K, requirement 3 the four-dimensional metrical determination, and requirement 1 the simultaneity section splitting off the spatial subspace r. It is obvious that r must be Euclidean. The space-time continuum R is Euclidean, because we take the Euclidean plane as the starting point; and since simultaneity in K means the same as in $\bar{\bar{K}}$, the split-off subspace r is the same. We note that in this case r has only two dimensions, because only the problem of the plane is treated; R has therefore three dimensions, but an indefinite metric.

The only transformations satisfying the three requirements are the trivial transformations

$$\left.\begin{aligned} x_1 &= \bar{\bar{x}}_1 \\ x_2 &= \bar{\bar{x}}_2 + \omega\bar{\bar{x}}_4\,, \\ x_4 &= \bar{\bar{x}}_4 \end{aligned}\right\} \qquad (108)$$

[54] Within the circle $r < 1/\omega$ this definition satisfies for any clocks the requirement of causality that a light signal travels in positive time, because (110) satisfies here the conditions of real systems; cf. § 46.

where x_1 and x_2 are the polar coordinates r and φ. The $\bar{\bar{g}}_{\mu\nu}$ of the plane for polar coordinates are

$$\left.\begin{matrix} 1 & 0 & 0 \\ 0 & \bar{\bar{x}}_1^{\,2} & 0 \\ 0 & 0 & -1. \end{matrix}\right\} \tag{109}$$

With (108) the $g_{\mu\nu}$ of the circular disk are obtained by means of the tensor transformation law $g_{\mu\nu} = \frac{\partial \bar{\bar{x}}_i}{\partial x_\mu} \frac{\partial \bar{\bar{x}}_k}{\partial x_\nu} \bar{\bar{g}}_{ik}$ as follows:

$$\left.\begin{matrix} 1 & 0 & 0 \\ 0 & x_1^{\,2} & \omega x_1^{\,2} \\ 0 & \omega x_1^{\,2} & (\omega^2 x_1^{\,2} - 1). \end{matrix}\right\} \tag{110}$$

There occurs a g_{24} so that the circular disk is not a static, but a stationary, system. In order that g_{44} be negative, the peripheral velocity must be $v = \omega x_1 < 1$, that is, that it be less than the velocity of light. Outside this region the rotating system is no longer real, that is, there are no stationary real points in this region.

The spatial geometry of the circular disk is given by $d\sigma^2 = g_{11}dx_1^2 + 2g_{12}dx_1dx_2 + g_{22}dx_2^2 = dx_1^2 + x_1^2dx_2^2$; it is Euclidean and identical with the spatial geometry of K, but not identical with the geometry of rigid rods on the disk. We demonstrate this fact as follows. We draw a circle around P (cf. fig. 14), whose radius is $d\sigma = 1$. If a rigid rod of the length 1 is laid through P in the radial direction, its characteristic length is $d\sigma = 1$, because in the radial direction the simultaneity of the circular disk corresponds to definition 26 of the characteristic length. Because of the adjustment of the clocks from the center the velocity of light is radially equal in both directions. If the rod lies tangentially, however, this is no longer the case. If two events P and Q happen simultaneously, defined in K, their $dx_4 = 0$ and because of $d\sigma = 1$ also $ds = 1$. But if at P a transition is made to the local orthogonal system K' while the rod is kept stationary, then the corresponding $dx_4' \neq 0$ and

$$ds^2 = 1 = d\sigma'^2 + g_{44}' \, dx_4'^2. \tag{111}$$

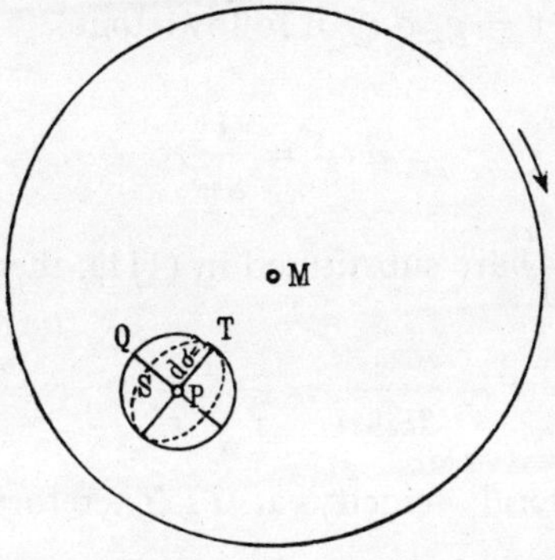

Fig. 14. The rotating circular disk. A circle with the radius $d\sigma = 1$ $(= PQ = PT)$ is described around P. If a rigid rod is rotated as diameter around P as the center, its end point describes the dotted ellipse.

Since g'_{44} is negative, $d\sigma'^2 > 1$. Since the tangentially lying rod in K' has the length 1—because this is its characteristic length—it is shorter than the distance PQ and equal to PS. If the rod is rotated around P, it describes the dotted curve.

In order to calculate this difference, we must transform into the local orthogonal system K'. This result is achieved by means of the linear transformation keeping the rod stationary:

$$\left.\begin{aligned} x_\alpha &= x'_\alpha \qquad \alpha = 1, 2. \\ x_4 &= x'_4 - p x'_2 \end{aligned}\right\} \tag{112}$$

In order to obtain $g'_{24} = 0$, we must set

$$p = \frac{g_{24}}{g_{44}}$$

as we discover by the transformation. g'_{14} will also be $= 0$ and $g'_{44} = g_{44}$. The $g'_{\mu\nu}$ become independent of time although this result is not necessary for our consideration.

For $dx_4 = 0$, then according to (112)

$$dx'_4 = p dx'_2 = p dx_2 \tag{113}$$

and from $d\sigma = 1 = g_{22}dx_2^{\,2}$ it follows that

$$dx_2^{\,2} = \frac{1}{g_{22}}. \tag{114}$$

If (113) and (114) are substituted in (111), then with $g'_{44} = g_{44}$

$$d\sigma'^2 = 1 - \frac{g_{24}^{\;2}}{g_{22}g_{44}} = \frac{1}{1 - v^2} \qquad v = \omega x_1. \tag{115}$$

v is the peripheral velocity at P. Therefore, the distance $PQ = \frac{1}{\sqrt{1 - v^2}} > 1$, measured by the rigid rod as unit. If conversely PQ is defined $= 1$ as in figure 14, the tangential rod has merely the length $\sqrt{1 - v^2}$. This result agrees with the calculation of the well-known circle paradox which receives this value from the Lorentz contraction. The dotted curve is an ellipse[55] whose radial semiaxis $= 1$ and whose tangential semiaxis $= \sqrt{1 - v^2}$.

This is the reason the geometry of the $d\sigma$ is Euclidean, whereas the geometry of the rigid rod is non-Euclidean. "Geometry of the rigid rod" means that the same rigid rod shall everywhere have the spatial length 1 while resting on the circular disk; in other words, the radii of the dotted curve are called equally long. The question may be raised whether this geometry can be combined with a definition of time so that it results in a $d\sigma$ geometry by way of a simultaneity section. This is possible; but for this purpose Einstein's definition of the metrical determination by ds^2 must be given up. The characteristic length of the rod would then not be $= 1$ in every orientation. Two events happening simultaneously at P and S, defined for K, would have a $ds = 1$, whereas according to definition 26 they have a

[55] This result is easily obtained if the given calculation is carried through for inclined positions of the rod. Then $d\sigma'^2 = \frac{1}{1 - v^2} - \frac{v^2}{1 - v^2} dx_1^{\,2}$ and this result leads to an ellipse.

$ds \neq 1$. If Einstein's metrical determination is retained, it will be impossible to make the geometry of the rod into a $d\sigma$ geometry. For this purpose the simultaneity of the whole system K would have to be defined everywhere according to definition 8, that is, it would have to be possible to make g_{24} vanish by means of a pure time transformation. The substitution of the $g_{\mu\nu}$ (110) of the circular disk in (107) shows that this is impossible. This equation results in a contradiction for $\alpha = 1$, $\beta = 2$. F. Kottler has demonstrated the same result in a different manner; he has also shown that the $d\sigma$ geometry becomes Euclidean.[56]

These considerations show at once that the round-trip axiom IV,2 does not hold on the circular disk. If light signals are sent from a clock along the periphery of the circle in both directions to a diametrically located clock and returned on equal [but not identical] paths, and if the clock is adjusted according to $t_2 = \frac{t_1 + t_3}{2}$, then different adjustments result for each direction. The stationary system differs from the static system in that the synchronism is not transitive. However, the ratio of the units of distant clocks is determined by light signals in the same way as in the static system and arises from the square root of the ratio of the g_{44}.

The space-time world of the rotating circular disk can be imagined easily. As mentioned above, the time may be imagined to be synchronized from the center according to $t_2 = \frac{t_1 + t_3}{2}$. It is not difficult to understand that any two clocks, compared with each other, will not satisfy this definition. We can think of a system in the classical theory of time moving uniformly

[56] *Phys. Zeitschr.*, vol. 22 (1921), p. 274. However, I do not believe that Kottler's "generalized Lorentz transformation" is fruitful. The complexity of the transformation is due to the fact that the central clock is not set up in the center. But if the simplest space-time measurement is desired, this great simplification must not be renounced. Furthermore, the relation between the $d\sigma$ geometry and the geometry of the rod is exhibited much more clearly in our presentation.

relative to the ether; according to the classical addition of velocities, the velocity of light along a certain distance is different in both directions. Even in this case, a special formula would have to be used for every comparison of two clocks by means of light signals. The transitivity or adjustability of any clocks according to the same rule is merely an advantageous property. But even without it time has an intuitive significance. The difference between the rotating circular disk and the moving system of the classical theory of time consists in the fact that although for the latter system no transitivity holds for the synchronization by means of light rays, there exists another method, namely, the application of the infinite limiting velocity by means of which transitivity can be achieved. No such method exists for the relativistic time of the rotating circular disk (§ 25). Under no circumstances can the existence of such a method be demanded as necessary for visualization.

The measurement of space is equally intuitive. Measurements are made with rigid rods that are laid down successively; a corrective factor dependent on the radius must be given to the tangential rod. The correction can be interpreted as compensation of the effect of a force that receives the properties of a metrical force (§ 18). The reason we do not set it equal to zero—although it would be possible—is our preference for setting the metrical force for the four-dimensional ds^2 equal to zero. In this case, obviously only one of the metrical forces, that of the three-dimensional rigid rod or that of the four-dimensional space-time interval, can be defined $= 0$.

We see that the metric in the nonstatic gravitational field can also be visualized clearly, although its presentation is somewhat more complicated. With the help of transformation (108), the metric (110), and figure 14 we have a complete command of the geometry of the rotating disk.

§ 45. *Stationary Fields. Continuation*

Since the synchronism is no longer transitive, the light-geometrical construction, too, fails in the stationary system.

Definition 10 of the equality of distances is no longer applicable. Figure 14 helps to make this clear. If light signals and definition 10 are applied, then PS would be $= PT$; but these distances have also in the $q_{\mu\nu}$ metrical determination an unequal $\overline{d\sigma} = \dfrac{d\sigma}{-g_{44}}$. The definition of time cannot be formulated in such a way that definition 10 would be valid for the $q_{\mu\nu}$; the reason for this fact is the same as for the geometry of the rigid rod. Axiom III does not hold anymore either, since Fermat's principle no longer applies.

This is the reason the direct construction of the metrical determination fails in the stationary case, and the method of § 38 has to be used. Only in special instances can the metrical determination be obtained through a transition to coordinate systems having a different state of motion and being static or even Euclidean. We used this method for the rotating circular disk. The rotating circular disk belongs also to a special class of stationary systems that can be reduced to a simpler type by means of a motion transformation, rather than by a mere time transformation.

For the matter axioms the conditions for the stationary and the static fields are the same. It is true, furthermore, that by the validity of the light axioms I, II, and IV, 1 a somewhat more general class is characterized than corresponds to the concept of stationary system, that class, namely, for which the $q_{\mu\nu}$ become stationary. If the stationary system is to be delimited further, a rigid rod or a clock must be added.

§ 46. *Real Systems*†

Systems with the following properties are to be understood by this name:

1. $g_{44} < 0$.
2. $d\sigma^2$ is to be a positive, definite, quadratic form, that is,

† See editor's reference notes, p. 204.

the following inequalities are to hold:

$$g_{11} > 0 \qquad \begin{vmatrix} g_{11} g_{12} \\ g_{21} g_{22} \end{vmatrix} > 0 \qquad \begin{vmatrix} g_{11} g_{12} g_{13} \\ g_{21} g_{22} g_{23} \\ g_{31} g_{32} g_{33} \end{vmatrix} > 0\,, \qquad (116)$$

or the inequalities resulting from these expressions by cyclical permutation of the indices 1, 2, 3.

3. The $g_{\mu\nu}$ are assumed to be real, continuous, and differentiable.

4. Only "apparent singularities" of the $g_{\mu\nu}$ may occur; if particular places occur where the $g_{\mu\nu}$ deviate from some of the requirements 1 to 3, it should be possible by means of a single-valued transformation to introduce a coordinate system in which at such a place requirements 1 to 3 are satisfied.

The first to point out such systems and their significance for causality was Hilbert;[57] he calls them "proper space-time coordinate systems." We prefer the shorter term "real systems," which will be justified by the following presentation. With Hilbert we call a metric satisfying requirements 3 and 4 a "regular metric."

We show first that in such systems a real point can be at rest at every point. For this purpose we have to demonstrate that by means of a real transformation one can go at every point of such a system to the local orthogonal system while the rod remains at rest and the temporal sequence is preserved. If this result can be achieved, the axioms of the special theory of relativity will hold infinitesimally for a stationary point of the system and according to axiom XI, 1 a real point must therefore be able to be at rest in this system.

We shall give this proof by asking for conditions which a system K must satisfy so that this transition may be possible. Let the transformations be

$$x_i = f_i(x_1' \ldots x_4')\,. \qquad (117)$$

[57] *Grundlagen der Physik II, Göttinger Nachr.* (1917), *Math.-phys. Kl.*, p. 53.

For the point under consideration the following expressions are to hold according to (85) and (88):

$$\frac{\partial x_\alpha}{\partial x'_4} = 0 \qquad \frac{\partial x_4}{\partial x'_4} > 0. \tag{118}$$

First, g'_{44} must be negative in the local orthogonal system K'; thus, because of (118),

$$g'_{44} = \frac{\partial x_i}{\partial x'_4}\frac{\partial x_k}{\partial x'_4} g_{ik} = \left(\frac{\partial x_4}{\partial x'_4}\right)^2 g_{44} < 0.$$

For a real transformation $\left(\frac{\partial x_4}{\partial x'_4}\right)^2 > 0$; therefore, $g_{44} < 0$. The first condition for K is therefore

$$g_{44} < 0. \tag{119}$$

The subsequent calculation is simplified by the following consideration. If it has been possible to eliminate the $g'_{\alpha 4}$ ($\alpha = 1, 2, 3$) locally by a transformation, and if the resulting $d\sigma$ is a positive, definite, quadratic form, then the transition from here to spatially orthogonal axes is possible by means of a purely spatial transformation. This result follows from the theorem of inertia for positive, definite, quadratic forms. We need merely look for the conditions that $g'_{\alpha 4} = 0$ and $d\sigma'$ be positive definite. We obtain

$$g'_{\alpha 4} = 0 = \frac{\partial x_i}{\partial x'_\alpha}\frac{\partial x_k}{\partial x'_4} g_{ik} = \frac{\partial x_\gamma}{\partial x'_\alpha}\frac{\partial x_4}{\partial x'_4} g_{\gamma 4} + \frac{\partial x_4}{\partial x'_\alpha}\frac{\partial x_4}{\partial x'_4} g_{44}, \qquad \alpha, \gamma = 1, 2, 3.$$

From this expression we obtain

$$\frac{\partial x_4}{\partial x'_\alpha} = -\frac{\partial x_\gamma}{\partial x'_\alpha}\cdot\frac{g_{\gamma 4}}{g_{44}}. \tag{120}$$

Furthermore,

$$g'_{\alpha\beta} = \frac{\partial x_i}{\partial x'_\alpha}\frac{\partial x_k}{\partial x'_\beta} g_{ik} = \frac{\partial x_\gamma}{\partial x'_\alpha}\frac{\partial x_\delta}{\partial x'_\beta} g_{\gamma\delta} + \frac{\partial x_4}{\partial x'_\alpha}\frac{\partial x_\gamma}{\partial x'_\beta} g_{4\gamma} + \frac{\partial x_\gamma}{\partial x'_\alpha}\cdot\frac{\partial x_4}{\partial x'_\beta} g_{4\gamma} + \frac{\partial x_4}{\partial x'_\alpha}\cdot\frac{\partial x_4}{\partial x'_\beta} g_{44},$$

$$\alpha, \beta, \gamma, \delta = 1, 2, 3,$$

and with (120) this becomes

$$g'_{\alpha\beta} = \frac{\partial x_\gamma}{\partial x'_\alpha} \cdot \frac{\partial x_\delta}{\partial x'_\beta} g_{\gamma\delta} - \frac{1}{g_{44}} \cdot \frac{\partial x_\gamma}{\partial x'_\alpha} \frac{\partial x_\delta}{\partial x'_\beta} \cdot g_{\gamma 4} g_{\delta 4} \,.$$

If we now form $d\sigma'^2 = g'_{\alpha\beta} dx'_\alpha dx'_\beta$, and set on the right-hand side because of (118)

$$dx'_\alpha = \frac{\partial x'_\alpha}{\partial x_\varepsilon} dx_\varepsilon \,, \qquad \alpha, \varepsilon = 1, 2, 3 \,,$$

and if we take into consideration that

$$\frac{\partial x'_\alpha}{\partial x_\varepsilon} \cdot \frac{\partial x_\gamma}{\partial x'_\alpha} = \frac{1}{D} \cdot D^\varepsilon_\alpha \cdot \frac{\partial x_\gamma}{\partial x'_\alpha} = \delta_{\varepsilon\gamma} \qquad (121)$$

(according to a well-known theorem of determinants[58] and because of (118)), then finally

$$d\sigma'^2 = d\sigma^2 - \frac{1}{g_{44}} d\gamma^2 \qquad d\gamma = g_{\alpha 4} \, dx_\alpha \,. \qquad (122)$$

We can now formulate the conditions for a positive, definite $d\sigma'^2$. Since $g_{44} < 0$ and $d\gamma$ is real, the last summand on the right-hand side is additive. If $d\sigma^2 > 0$, then $d\sigma'^2 > 0$. But this is not the necessary condition for a positive $d\sigma'^2$; even for a negative $d\sigma^2$ one can go to the local orthogonal system (whose $d\sigma'^2 > 0$) by means of real transformations while the rod is kept stationary. The necessary condition that the $g_{\mu\nu}$ in K must satisfy for this purpose is obtained if the right-hand side of (122) is written in quadratic form:

$$d\sigma'^2 = p_{\alpha\beta} \, dx_\alpha \, dx_\beta > 0; \; p_{\alpha\beta} = g_{\alpha\beta} - \frac{1}{g_{44}} g_{\alpha 4} g_{\beta 4} \,. \qquad (123)$$

The coefficients $p_{\alpha\beta}$ must then satisfy the familiar conditions of positive, definite, quadratic forms; they will be the conditions

[58] Cf. p. 137 for the significance of D and D^ε_α, and p. 83 n. concerning $\delta_{\varepsilon\gamma}$.

for the g_{ik}. Therefore, the following inequalities must hold:

$$p_{11} > 0 \qquad \begin{vmatrix} p_{11} p_{12} \\ p_{21} p_{22} \end{vmatrix} > 0 \qquad \begin{vmatrix} p_{11} p_{12} p_{13} \\ p_{21} p_{22} p_{23} \\ p_{31} p_{32} p_{33} \end{vmatrix} > 0\,, \qquad (124)$$

or those inequalities resulting from (124) by means of cyclical commutation of the indices 1, 2, 3.

If the system K is to admit at every point at any time the transition to the local orthogonal system upon a real transformation while the temporal sequence is preserved and the rod remains stationary, then (119) and (124) must hold at every point and at any time.

We shall now show that a positive $d\sigma^2$ can be obtained everywhere, that is, for the system as a whole, from a system K of such properties by means of a mere real time transformation. If we go from K to a system K'' by means of the transformation

$$\left.\begin{aligned} x_\alpha &= x''_\alpha \\ x_4 &= f(x''_1 \ldots x''_4)\,, \end{aligned}\right\} \qquad (125)$$

then

$$g''_{\alpha\beta} = \frac{\partial x_i}{\partial x''_\alpha} \cdot \frac{\partial x_k}{\partial x''_\beta}\, g_{ik} = g_{\alpha\beta} + \frac{\partial x_4}{\partial x''_\alpha}\, g_{4\beta} + \frac{\partial x_4}{\partial x''_\beta}\, g_{4\alpha} + \frac{\partial x_4}{\partial x''_\alpha} \cdot \frac{\partial x_4}{\partial x''_\beta}\, g_{44}\,.$$

This can be transformed into

$$\begin{aligned} g''_{\alpha\beta} &= p_{\alpha\beta} - r_{\alpha\beta} \\ r_{\alpha\beta} &= r_\alpha r_\beta \qquad r_\alpha = \sqrt{-g_{44}} \cdot \left(\frac{\partial x_4}{\partial x''_\alpha} + \frac{g_{4\alpha}}{g_{44}}\right), \end{aligned} \qquad (126)$$

so that because of $dx_\alpha = dx''_\alpha$ we obtain

$$d\sigma''^2 = g''_{\alpha\beta}\, dx''_\alpha\, dx''_\beta = p_{\alpha\beta}\, dx_\alpha\, dx_\beta - r_{\alpha\beta}\, dx_\alpha\, dx_\beta\,. \qquad (127)$$

Here $p_{\alpha\beta} dx_\alpha dx_\beta$ is a positive, definite quadratic form according to (124); but $r_{\alpha\beta} dx_\alpha dx_\beta$ is also one, because it is $= (r_\alpha dx_\alpha)^2$ or a complete square. ($\sqrt{-g_{44}}$ is real.) Thus on the right-hand side the difference of two positive terms occurs. It is obvious that (124) is a necessary condition for a positive $d\sigma''^2$. It is also a sufficient condition. By means of the choice of the coefficients

r_α it can be arranged that, for each choice of dx_α, the form $r_{\alpha\beta}dx_\alpha dx_\beta$ will become smaller than the form $p_{\alpha\beta}dx_\alpha dx_\beta$ with the same dx_α. If conditions (124) of the coefficients of positive, definite, quadratic forms are written down for $g''_{\alpha\beta} = p_{\alpha\beta} - r_{\alpha\beta}$, then three inequalities result for the three magnitudes r_α, and with (126) three inequalities are obtained for the $\frac{\partial x_4}{\partial x''_\alpha}$ of the form

$$\frac{\partial x_4}{\partial x''_\alpha} < t_\alpha, \tag{128a}$$

where the t_α are given functions of the four coordinates that are constructed from the ten functions $g_{\mu\nu}$. If transformation (125) is chosen in correspondence to these inequalities, a real system is obtained in K''.

It can be shown that the t_α are always real functions if the $p_{\alpha\beta}$ satisfy the inequalities (124). I could not demonstrate rigorously, however, that the inequalities (128*a*) always admit real solutions (125), that is, for any given form of the t_α. Yet it seems very plausible, because the Cauchy-Riemann integrability conditions do not pose any difficulties since (128*a*) represents only an inequality. In any event, one can say that if there are cases in which (128*a*) cannot be satisfied they are rare exceptions. We shall therefore no longer consider this possibility.

No requirement corresponding to (128*a*) is made concerning the derivative $\frac{\partial x_4}{\partial x''_4}$. We may therefore stipulate

$$\frac{\partial x_4}{\partial x''_4} > 0 \tag{128b}$$

This stipulation has also the result that the determinant of transformation (125) does not vanish anywhere; otherwise, this determinant would vanish for $\frac{\partial x_4}{\partial x''_4} = 0$ because of (85). Condition (128*b*) signifies the conservation of temporal sequence (in this case it is also symmetrical because of the identity transformation of the spatial coordinates, that is, because of (85)). Transformation (125), which transforms the most general system in which stationary real points occur into a real system, constitutes merely a change of the definition of simultaneity and of the metric of the linear temporal sequence at the individual real points.

However, it is generally impossible to eliminate the $g_{\alpha 4}$ in a system with property (124) by a mere time transformation. This result would require $r_\alpha = 0$, and this leads with the formation of $\frac{\partial^2 x_4}{\partial x''_\alpha \partial x''_\beta}$ to the integrability condition (three equations, $\alpha \neq \beta$):

$$\frac{g_{\beta 4}}{g_{44}} \cdot \frac{\partial}{\partial x_4}\left(\frac{g_{\alpha 4}}{g_{44}}\right) - \frac{\partial}{\partial x_\beta}\left(\frac{g_{\alpha 4}}{g_{44}}\right) = \frac{g_{\alpha 4}}{g_{44}} \cdot \frac{\partial}{\partial x_4}\left(\frac{g_{\beta 4}}{g_{44}}\right) - \frac{\partial}{\partial x_\alpha}\left(\frac{g_{\beta 4}}{g_{44}}\right). \quad (129)$$

This is an internal relation of the g_{i4} and their derivatives which must be satisfied in addition to (124); it means that in the given real system a definition of time can be carried through that becomes locally identical at every point with the simultaneity of the special theory of relativity.

We find therefore that the particular systems in which stationary real points exist form a somewhat more general class than the real systems. The additional systems can be obtained from the real systems alone by a change of the definition of time. We shall call them therefore *real systems with a deviating definition* of time. Together they are characterized by the invariant conditions (119) and (124).

At the same time, we have demonstrated that a real point can be stationary at every point of a real system.

§ 47. *The Validity of Axioms I and II in Real Systems*

We shall now investigate which axioms are still valid in real systems. Axiom IV, 1 is no longer valid, because the $g_{\mu\nu}$

are generally temporally variable. Axiom III is no longer valid either (cf. § 45). We shall show, however, that axioms I and II are still valid.

For a light signal $ds^2 = 0$, and so we obtain

$$dx_4 = -\frac{d\gamma}{g_{44}} + \sqrt{\frac{d\gamma^2}{g_{44}^2} - \frac{d\sigma^2}{g_{44}}}. \qquad (130a)$$

Because of $g_{44} < 0$ and $d\sigma^2 > 0$, the root is real, since also $d\gamma^2 > 0$. But we can infer more. Although the sign of the root is arbitrary, it may be stipulated only once; therefore, we may make it positive. This stipulation constitutes merely a decision concerning which direction of time is to be called positive. However, the preceding term $-\frac{d\gamma}{g_{44}}$ does not have a definite sign; the sign of $d\gamma$ is reversed when the signs of the dx_α are reversed, that is, when the light signal along the same spatial line element $d\sigma$ is sent in the opposite direction. Since $\frac{-d\sigma^2}{g_{44}}$ is always positive, the root is greater than $-\frac{d\gamma}{g_{44}}$ in any case; thus dx_4 will be positive for any choice of the dx_α. A light signal will therefore always receive positive time differences on its path from the clocks it passes; the definition of time of the real system corresponds to theorem 9. Properties 1 to 4 of the real system are the necessary and sufficient conditions for it.

In real systems with a deviating definition of time this requirement is no longer satisfied. In this case it is possible that a temporally advancing light signal will have negative time differences. We notice, however, if (130*a*) is written in the form

$$\begin{aligned} dx_4 &= -\frac{d\gamma}{g_{44}} + \frac{1}{\sqrt{-g_{44}}} \cdot \sqrt{d\sigma^2 - \frac{d\gamma^2}{g_{44}}} \\ &= -\frac{d\gamma}{g_{44}} + \frac{1}{\sqrt{-g_{44}}} \cdot \sqrt{p_{\alpha\beta}\, dx_\alpha\, dx_\beta} \qquad (130b) \end{aligned}$$

that the characterization (124) of these systems says explicitly that the root will always be real. Thus in real systems with a deviating definition of time, the time difference for light signals will at least always be real.

Axiom I, 1 holds for real systems because of the above-mentioned peculiarity of the definition of time. For every light signal returning to its spatial starting point P, the $\int dx_4$ taken along the spatial path has a positive value, and because of the continuity of the definition of time the time difference between return and departure of the signal, determined from P, is also positive. In real systems there are therefore no timelike closed world lines.†

Furthermore, axioms I, 2 and I, 3 are satisfied. At P the special theory of relativity is valid infinitesimally, since it is a real point; this fact is sufficient for the validity of the axioms. Although the special theory of relativity holds temporally only in infinitesimal regions, these axioms must hold integrally with respect to time, because the special theory of relativity holds at any moment at P.

Axioms II, 1 and II, 2 also hold. On the one hand, these axioms depend upon the fact that no temporal singularities occur, that is, that the time sequence does not suddenly begin at any point; on the other hand, the axioms say that the world does not consist of separate parts. These facts are included in the definition of the real system by the requirement of the continuity of the $g_{\mu\nu}$.

We shall now discuss axioms II, 3 and II, 4; theorem 4 is equivalent to them and it suffices therefore to consider this theorem. This consideration meets with difficulties because axiom III no longer holds. In general fields, the first signal need not even be a reflected signal but merely a limit of such signals. We can divide our analysis into two steps. First, we shall consider only direct signals, that is, exclude other signals

† This proof of the satisfaction of axiom I, 1 in real systems is shown to be invalid in *PST*, pp. 272–273: "The proof overlooks the fact that the existence of closed world lines demands only an 'apparent singularity' (*AR*, p. 80) of the coordinate system."—*Ed.*

from axioms I and II,2; then axiom III holds for this class of signals. We shall investigate whether theorem 4 also holds for direct signals. Only after we have proved this shall we advance to the second step and show that theorem 4 holds for all signals.

Let us start with the first step. Because $dx_4 > 0$, the wave surfaces emanating from a point P (whose shape will be defined by means of the definition of simultaneity) are stratified in such a way that every outer wave surface encloses the inner one completely without touching it. It follows that no direct signal can arrive earlier at a distant point P_1 than a signal that departed earlier from P, because if one signal overtook the other, then the special theory of relativity would no longer hold at that particular point.

We must assume for this purpose that the topological relations of the three-dimensional space do not change temporally. This problem must now be investigated. If this assumption is not made, a signal leaving P later and arriving at P_1 earlier without overtaking the first signal is possible. Because of a topological change the outer wave surfaces would become surfaces that cannot be contracted to a point. Figure 15 is a two-dimensional analogue to curves on a torus. If the surface is still open, that is, if it is not a torus, then the closed wave lines around P spread as a family of curves which can be contracted to P. When the signal, which must be thought of as a

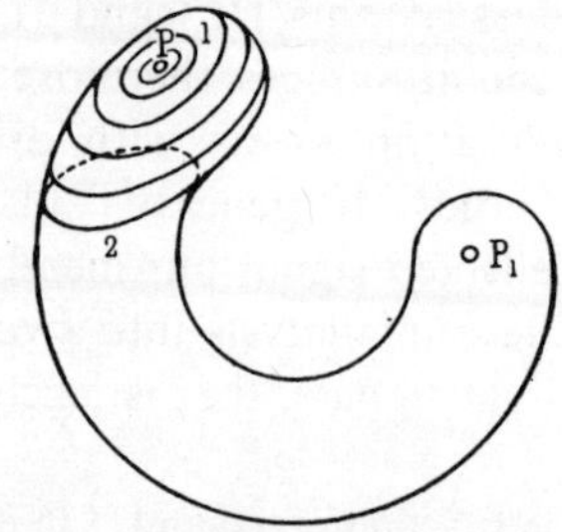

Fig. 15. A two-dimensional analogue. After the surface (still drawn in an open fashion) has become a closed torus, a second path from P to P_1 exists for the toroidal waves.

single impulse, has arrived at position 2, the surface may become a closed torus so that now the signal forms a closed curve that cannot be contracted to a point. If now a signal travels from P to P_1, it has a much shorter, direct path and will arrive earlier than the first without overtaking it on the way.

However, at that place where the ends of the torus first touch each other, the special theory of relativity would not hold at this moment; this point event would be a genuine singularity. Every change of the topological relations of three-dimensional space in a real system would be a genuine singularity, because in this case real points would suddenly appear in the infinitesimal neighborhood of a point P_x that previously had not been located in its neighborhood. Under these circumstances, for instance, axiom IV, 1 would not hold infinitesimally for these real points. The metric of the field would express the singularity by the fact that the $g_{\mu\nu}$ would be discontinuous and not differentiable at that particular place and that this irregularity could not be eliminated by means of a unique transformation; it would be clear that the irregularity did not arise merely from an unsuitable choice of the coordinate system. Since we included the continuity and differentiability of the $g_{\mu\nu}$ in the definition of the real system, we have excluded singularities, and the case of a change of the spatial topology is out of the question.

We do not here say anything about the topological relations themselves; even in real systems, any kind of topology can exist. All that is excluded is the temporal change of the topological relations of the R_3. This can be imagined as follows. The specific definition of time in the real system has the consequence that the world naturally divides into space and time. The simultaneity cross sections $x_4 = \textit{constant}$ cut out subspaces R_3 in such a way that these subspaces do not intersect; the world is layered cylindrically. If the topological relations of the R_3 change temporally, this would mean that between two subspaces R_3 that correspond to different values $x_4^{(1)}$ and $x_4^{(2)}$ no unique transformation exists. Since each of these R_3 consists of real points, this means that at a certain time x_4 lying between $x_4^{(1)}$ and $x_4^{(2)}$ the neighborhood relations of the real points have changed in a discontinuous manner at some point in space. This change constitutes a violation of axioms IV, 1 and X of the special theory of relativity in infinitesimal regions. This does not mean that such genuine singularities cannot occur in nature: they are excluded only if the special theory of relativity holds without exception in infinitesimal regions. It should be noted, furthermore, that even this axiom does not suffice for the justification of the assertion; it is essential that the system be a real one, that it can be arranged in temporal layers, and that the R_3 consist of real points. Only for real points does the discontinuous change of the neighborhood relations constitute a genuine singularity; for fictitious points it can be produced by the coordinate system alone without a violation of the special theory of relativity at this space point.

Thus we have shown that the transfer of the time sequence according to theorem 4 holds in real systems for direct signals. It is valid for first signals, too, as we are going to show by means of the second step of our analysis. The inference used until now is not applicable here, because first signals can be reflected on their way and are therefore capable of overtaking other waves of the same center without violating the special theory of relativity. But we can prove theorem 4 by a different method. We consider the two partial assertions, axioms II, 3 and II, 4, of this theorem separately. We assume first, in contradiction to axiom II, 4, that a first signal L_2 exists leaving P later than the first signal L and arriving at P' simultaneously with L. Then a signal L_1 can be sent from P simultaneously with L traveling the same path W as L_2, thus always touching points with the same spatial coordinates. If the natural definition of time is carried through, then the various positions of L_2 and L_1 on the path W can be so ordered with respect to time that L_2 always lags only a little behind L_1. On the same path, that is, by means of the same reflecting mirrors, L_1 cannot overtake L_2; otherwise, the special theory of relativity would be violated at that particular point. Therefore, L_1 arrives at P' before L_2, and L is not a first signal. It is possible that L_1 is not a first signal either, because at this time a signal will reach P' perhaps faster along another path, whereas at a later time W is the path of the first signal L_2. But then a first signal exists that reaches P' before L_2. Thus the assumption of the invalidity of axiom II,4 leads to a contradiction and, therefore, this axiom must hold.[59] This inference presupposes, too, that the topological relations of three-dimensional space do not change with respect to time, since it is assumed that the same path W is possible for L_1 as well as for L_2. The validity of axiom II,3 is inferred in a similar fashion. If a first signal is sent from P later than L, it will also arrive later at P_1 according to our demonstration. Because of the continuity of the $g_{\mu\nu}$ the difference between the times of arrival at P_1 must approach 0 if the difference between the times of departure from P approaches 0.

Axiom II,5 also holds; this axiom formulates the existence of a limiting velocity for arbitrary gravitational fields. Axiom II,6, or at least theorem 8*a*, will generally hold, too, because in real systems the time definition satisfies (1) without the equality sign, and theorem 8*a* or axiom II,6 is a presupposition of this property. Neither of them are necessary presuppositions (cf. p. 38), and there may be special cases where theorem 8*a* and axiom II,6 do not apply and the equality sign in (1) will be eliminated by a variation of ε in definition 2.

The adjustment of time in real systems may be imagined in such a way that all clocks are synchronized from a central clock according to definition 2 (aside from the occasional exceptions mentioned above where ε must be varied). The choice $\varepsilon = \frac{1}{2}$ has no advantage any longer, because the synchronism will not be transitive anyway. Axioms I and II, definitions 1 to 4, and theorems 1 to 10 contain the integral

---

[59] For the case that the first signal itself is no signal, this inference is not strict. We accept this imprecision, however, because the case that axiom II,4 is not valid for such first signals is presumably very rare. Until now no theoretical investigations exist concerning the first signal in a general gravitational field in which Fermat's principle no longer holds.

properties of the resulting time; they mean that space and time obtain a minimum of real significance. The most important property is the validity of the requirement of causality that all causal processes occur in the same direction of time. If only those coordinate transformations are considered that transform a real system into another real system, then—as Hilbert was the first to emphasize—a causal sequence will never be transformed into a sequence of simultaneous events and a spatial distance will never be transformed into a time sequence. On the other hand, the failure of axioms III, IV, and V means that metrical characteristics no longer exist. In real systems the concepts of spatial distance, temporal process at one point, temporal process between different points, and causal sequence retain therefore the intuitive significance obtained from the special theory of relativity for these concepts if all metrical determinations are renounced. The topological character of the order of time is identical with that of the special theory of relativity even in these very general gravitational fields.

§ 48. *The Most General Admissible Systems*†

One might contend (as Hilbert, *op. cit.*, n. 57, seems to assume) that all gravitational fields occurring in nature must always have the property of permitting the choice of a coordinate system that is real everywhere. The question arises, however, whether this property can be derived from the differential principle (axiom XI) which we presupposed for the general theory of relativity, or whether such an assertion depends on additional axioms.

The *most general admissible systems*, that is, those compatible with the validity of the special theory of relativity in infinitesimal regions, are characterized by the following properties:

1. The $g_{\mu\nu}$ are the coefficients of an indefinite quadratic

† See editor's reference notes, p. 204.

form of the inertia index 1; the following inequalities apply:

$$g_{11} > 0 \quad \begin{vmatrix} g_{11} g_{12} \\ g_{21} g_{22} \end{vmatrix} > 0 \quad \begin{vmatrix} g_{11} g_{12} g_{13} \\ g_{21} g_{22} g_{23} \\ g_{31} g_{32} g_{33} \end{vmatrix} > 0 \quad \begin{vmatrix} g_{11} g_{12} g_{13} g_{14} \\ g_{21} g_{22} g_{23} g_{24} \\ g_{31} g_{32} g_{33} g_{34} \\ g_{41} g_{42} g_{43} g_{44} \end{vmatrix} < 0 \quad (131)$$

as well as those inequalities originating from them by means of cyclical permutation of the indices 1, 2, 3, and 4.

2. The functions $g_{\mu\nu}$ are continuous and differentiable.

3. Only apparent singularities of the $g_{\mu\nu}$ occur, that is, deviations from 1 and 2 that can be eliminated by a biunique transformation into another coordinate system.

With respect to the first requirement, these systems are different from the real systems for which the indefinite form "breaks up"; these systems are therefore not always real. Thus the coordinates of the rotating circular disk are an admissible system outside the circle $r = 1/\omega$. In this domain they admit a biunique transformation into stationary Cartesian coordinates of the plane (the transformation determinant is x_1^2); but this is a transformation of motion and because of the positive g_{44} such a transformation is not possible while the rod remains stationary. Outside the circle there are no stationary real points in this system. In such a system space and time can no longer be imagined, but it is still a system biuniquely and continuously coordinated to the coincidences of reality, because the transition to the special theory of relativity by means of a biunique transformation is infinitesimally possible at every point. Of course, axioms I and II no longer apply to the fictitious, stationary points of such a system.

The question arises whether or not the deviating relations of such a system result merely from an unsuitable choice of coordinates, that is, whether or not axioms I and II would apply even in this case to real points everywhere. This question asks whether the most general admissible system can always be transformed into a real system for the world as a whole by means of a biunique transformation. This transformation must be biunique, because the coincidence must not be destroyed; otherwise one could not infer from the validity of axioms I and II in the introduced real system the validity for the real points of the original system.

Because of mathematical difficulties this question cannot be answered definitely today. It is true that a continuous transformation into a real system always exists—the $g_{\alpha 4}$ can always be eliminated and the g_{44} be made negative ("Gaussian coordinates," according to Hilbert, *op. cit.*, n. 57); according to the theorem of inertia of the quadratic forms, a real system results, but it is uncertain whether this transformation can always be chosen to be biunique.

It can be shown, however, that around every point of a most general admissible system a finite region can be delimited in such a way that a biunique transformation into a real system is possible. We reason as follows. When Gaussian coordinates have been established, they have the property that at every point (aside from singularities) the transition to the local orthogonal system by means of a transformation with a finite determinant is possible (this fact follows from the theory of quadratic forms). A transition with a finite determinant to the local orthogonal system is also possible from the most general admissible coordinate system and, therefore, the determinant of the transformation from the most general admissible system to the Gaussian system consists of the product of two such determinants and is therefore finite. It follows that around every world point a finite domain can be delimited (restricted by singularities) in which transformations are biunique.†

The inference from this result to the world as a whole is precluded for the time being because of the difficulty of characterizing the biuniqueness of a transformation on the basis of differential conditions. We are dealing with a problem of differential geometry in large domains. But its physical importance must not be overestimated. If the solution is positive, it will certainly contain the presupposition that no genuine

---

† See editor's reference notes, p. 204.

singularities will occur. But whether or not the special theory of relativity actually holds at every point of the world at every moment is a question which the physicist will hardly dare to decide so long as subatomic processes are still as theoretically inaccessible as today. It will therefore be expedient to restrict the statements concerning integral properties to finite domains.

For the same reason, the second method of giving to the world as a whole the character of a real system, for instance, by means of additional new axioms of causality, does not seem to be sufficiently justified either. It goes beyond the right of extrapolation to infer statements about the world as a whole from certain ideas to which we have become accustomed in our limited field of research.

We may therefore consider the properties of the real system, axioms I and II, as the *topological properties of the order of time in finite domains of the most general gravitational fields.* Signals not going beyond a certain finite space-time domain (that can be indicated in the environment of every world point) are not closed with respect to time, and an order of time is possible in which all signals are assigned continuous positive time intervals. Restricting ourselves to finite domains in the environment of every world point, we may say therefore that space and time in the general theory of relativity mean the same as in the special theory although without any metric and transitive simultaneity. The topological properties turn out to be more constant than the metrical ones, and the transition from the special theory to the general one represents merely a renunciation of metrical characteristics, while the fundamental, topological character of space and time remains the same. The distinction of rigid or uniformly moving coordinate systems is a metrical determination and therefore disappears in the general theory. At the same time, the concept of the relative rest of distant points loses its absolute significance and it depends on the choice of the coordinate system whether two real points are to be regarded as being at rest or moving relative to each other. We have operated several times with such systems, and have given in definition 23 a definition of the coordinate system that contains the general case. The relativization of relative motion expresses perhaps most clearly that the general theory has given up the "rigid space" of the special theory. Yet, in spite of all these renunciations, the difference between space and time is preserved. In finite domains special spatial coordinate systems exist which consist everywhere of real points and for which a consistent order of time in terms of causal order can be carried through. Although there is no necessity to choose the coordinates according to the rules of the real system—any four biunique functions of such preferred coordinates achieve the same result—an essential property of the space-time continuum is expressed in the possibility of such a simple choice of coordinates. Whoever has realized that a topological structure restricted to finite domains can be called an object of visualization to a much larger degree than the metric and all properties of the world as a whole, will see in this result, namely, in the *topological separability of the world into space and time*, a means of visualizing even the physical world of the general theory of relativity.

List of Axioms

Other Systems

General Theory

List of Definitions

General Theory

List of Theorems

Reference Notes

(*Page number at beginning of each note indicates page in text to which note refers and which is marked by dagger*)

Page xiii. MPS, p. 41, refers to *AR* for a more detailed analysis of Einsteinian kinematics which "rests on the hypothesis that light geometry is identical with the geometry of rigid rods and natural clocks. This hypothesis is asserted by a special group of axioms."

Page xiv. Cf. *PST*, p. xv. In his introduction to *PST*, Reichenbach explains that he has omitted detailed mathematical computations, because a considerable part of the necessary mathematical work had been completed in *AR*.

Page 12. *MPS*, p. 62 n., refers to *AR*, § 2, which contains for the first time Reichenbach's distinction between descriptive and inductive simplicity. The reference is used for a correction of Leibniz' view that the Copernican system is the simplest and, therefore, true. See also *MPS*, p. 123.

Page 14. *MPS*, p. 51 n., characterizes *AR* as a continuation of Leibniz' ideas, although at the time of writing the author had not been aware of this connection, since he was not then acquainted with the relevant passages of Leibniz.

MPS, pp. 53, 53 n., refers to the similarity between Leibniz' construction of the theory of space and time and that of the author. "Leibniz starts with the order of time and considers the order of space only after he has given a determination of simultaneity, a procedure which turned out to be fruitful for later axiomatization of the theory of space and time." Reichenbach explains that in *AR* the concept *spatially nearer than* is reduced to the concept *temporally earlier than* by means of a special kind of causal sequence: first signals.

Page 15. *DT*, p. 25, refers to the causal theory of time developed in *AR*. Whereas in *AR* the relation of cause and effect was assumed as a primitive term, *DT* treats it no longer as primitive but defines it in terms of more elementary relations.

Page 16. Cf. *PST*, p. 286 n., for this distinction.

Page 25. Cf. *PST*, p. 165 n. The construction of the space-time metric in *PST* presupposes the presentation given in *AR*.

Cf. also *DT*, p. 20. The theory of the metrical properties of time developed in *AR* is presupposed in *DT*.

Page 29. Cf. *PST*, p. 137 n. Instead of the reference to *AR*, p. 133, the reference should be to *AR*, p. 22 (of the German edition). The reference for the English edition of *AR* is p. 29.

For the author's later views on the issue of causality and irreversible processes as a definition of temporal order, see *PST*, pp. 138–139; "Ziele und Wege der physikalischen Erkenntnis," *Handbuch der Physik*, Vol. IV (1929), pp. 53, 59–60, 64, 65; "The Philosophical Significance of the Theory of Relativity," in *Albert Einstein: Philosopher-Scientist*, ed. P. A. Schilpp (Evanston, Ill.: The Library of Living Philosophers, Inc., 1949); *DT*, p. 198 n.; "Les Fondements Logiques de la Mécanique des Quanta," *Annales de l'Institut Henri Poincaré*, Vol. XIII (1953), pp. 137–138; *The Rise of Scientific Philosophy* (Berkeley and Los Angeles: University of California Press, 1951), pp. 148, 159.

Page 32. Cf. *EP*, p. 128, n. 5.

Page 36. Cf. *PST*, p. 168, which presupposes the above proof.

Page 40. Cf. *PST*, p. 124 n., which presupposes *AR*, § 8, for a treatment of the comparison of neighboring events.

Page 65. The detailed discussions of this example in *AR*, pp. 65, 162, are presupposed in *PST*, p. 222. There, reference is also made to the example of a system of radially expanding points given in *AR*, p. 78, in which the light geometry is Euclidean. Cf. *PST*, p. 222 n.

Page 66. Cf. *PST*, p. 208, n. 1, which refers to *AR*, § § 14–15.

Page 76. *PST* presupposes the calculations given in *AR*, § 16. Cf. *PST*, p. 173 n., for a reference to a correction.

Page 81. Cf. note on p. 55.

Page 83. Cf. *PST*, p. 175, equation (6), which presupposes the exact presentation given in the following section.

Page 88. In *PST*, p. 13, Reichenbach replaced the terms "metrical forces" by "universal forces," and "physical forces" by "differential forces" in order to avoid confusion.

Page 94. *PST*, § 31, presupposes *AR*, § § 21, 24.

Page 97. Cf. *EP*, p. 128, n. 6.

Page 109. Cf. *PST*, p. 192: "The relativistic hypothesis has been confirmed by observation of traveling atomic clocks which in fact show Einstein's retardation." Reference is made to the experiments of H. Ives on the transverse Doppler effect in cathode rays. *AR*, § § 23–24, is presupposed.

Page 117. Cf. *PST*, p. 149.

Page 144. Cf. *PST*, p. 230, n. 1.

Page 172. Cf. *PST*, p. 262 n.

Page 179. Cf. *PST*, p. 239 n. and § 45.

Page 192. Cf. *PST*, pp. 263 n., 273.

Page 194. Cf. *PST*, p. 285.

Index